Engineering Design for Technicians

Full Coverage of
TEC Engineering Design III

Barry Hawkes AMIED
Lecturer in Engineering Design and Mathematics
Canterbury College of Technology
Visiting Lecturer, University of Kent

Ray Abinett MIED
Principal Design Engineer, Lucas CAV Ltd
Lecturer in Engineering Design, Mid-Kent College
of Higher and Further Education, Chatham

Pitr

Pitman Education Limited
39 Parker Street, London WC2B 5PB

Associated Companies
Pitman Publishing New Zealand Ltd, Wellington
Pitman Publishing Pty Ltd, Melbourne

First published in Great Britain 1981

Printed and bound in Great Britain
at The Pitman Press, Bath

ISBN 0 273 01675 X

Contents

Preface

Engineering design appreciation is an important part of the engineers training. Whether a technician, graduate or polytechnic student the need for a basic understanding of elements in design has become more important, especially with the advent of the Technician Education Council.

This book has been prepared with a view to satisfying the Technician Level III Engineering Drawing and Design courses now being taught in Colleges and Polytechnics.

The first part of the book is concerned with appreciating how materials affect design thinking from physical aspects to the selection of the right material for a known load condition. Consideration has been given to basic design principles of machine elements and the safety aspects of protecting the operator from injury.

Logical steps in the design of a product is an essential part of design and to emphasise this aspect objectively the systematic process has been covered in some detail, showing how important it is to evaluate more than one idea to obtain the best solution to a problem.

Finally man and his relationship to the machine has been covered in the last chapter under the heading of Ergonomics, and shows the basic Control Loop system which dictates the operation of all machines.

To test the students learning ability and give emphasis to the practicalities of design, several design assignments have been included.

Essential charts and tables necessary in solving set design problems have been included, and also to emphasise the importance of these in engineering design.

The book sets out to show how design is controlled by all the above-mentioned aspects, and is intended to show the student that design is not just a matter of putting pencil to paper but also the application of defined engineering principles.

R. E. Abinett
B. Hawkes
Rochester/Maidstone 1981

For Barbara, Suzanne, Tracy, Jane,
and Gill and Georgia

1 Properties of Materials

With today's wealth of information on materials, it can be a demanding task to select the right material for a particular application. Many aspects have to be considered in designing a product, for example the environment in which the product is to operate and the ability of the materials to withstand the design loads. The physical and mechanical aspects of materials can have a far-reaching effect on the operation and durability of machine parts. In fact, the material's properties determines its selection or rejection.

Some of the more important properties are presented briefly in this chapter. However, there are many others and it is the designer's responsibility to satisfy the design specification by ensuring that the material chosen will have the necessary physical and mechanical properties.

Table 1.1 Densities of common engineering materials

Material	*Density* kg/m³
Aluminium	2568
Brass	8203/8604
Copper	8957
Cast iron	7223
Nickel	8298/9198
Solder	8909/8347
Tin	7304
Zinc	7062/7175

1.1 Density

The **density** of a substance is the mass of that substance per unit volume. For water at a temperature of 15.5°C (60°F), the density is 1000 kg/m^3.

The density of steel varies, depending on the carbon content, but for most design calculations it can be taken as 7850 kg/m^3.

The densities of some other popular engineering materials are given in Table 1.1.

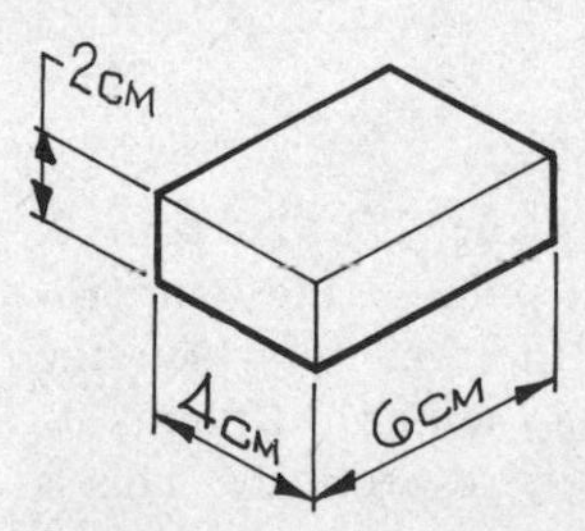

Fig. 1.1

Example

Compare the mass of a component (Fig. 1.1) made from steel and from aluminium.

It can be seen that, for the same volumetric shape, the steel component is approximately three times the mass of one made from aluminium. In design terms, the mass saving can be advantageous where load bearing is a criteria.

1.2 Thermal Conductivity

The **thermal conductivity** of a substance is the ability of that substance to allow the passage of heat through its body. The rate at which this occurs depends on the density of the mass of the substance.

The unit of measurement is watt/metre°C (or watt/metre°K).

Table 1.2 Thermal conductivites of some common engineering materials

Material	*Thermal conductivity* W/m°C
Aluminium	236
Brass	106
Copper	403
Cast iron	835
Nickel	94
Tin	68
Zinc(cast)	117
Magnesium	157
Steel(carbon)	50
Steel(stainless)	245

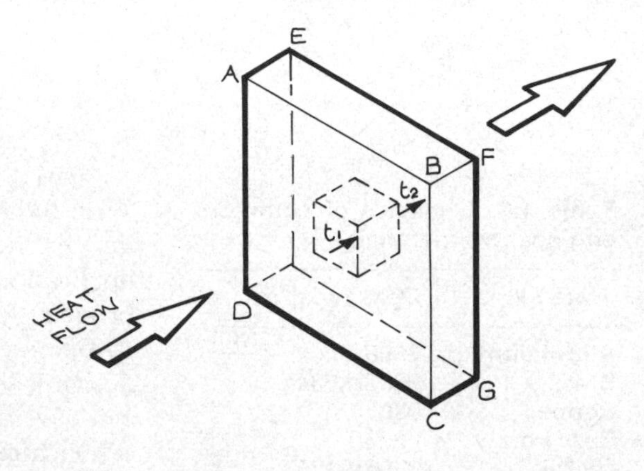

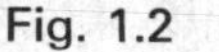
Fig. 1.2

It can be related to the heat flow, by considering conduction through a very large plate (see Fig. 1.2). The heat flow is from A to E in the direction of the arrows. It is assumed that the plate is large enough to ignore the heat escaping from the edges.

Consider a cube of one centimetre in the middle. Heat is flowing through this and, because of the density of the material, there is a change in temperature from one side to the other, i.e. $t_1 \rightarrow t_2$.

The conductivity K of the substance can therefore be found by dividing the quantity of heat Q flowing per second through the unit cube by the temperature difference $t_1 - t_2$ of the opposite faces of the cube, i.e.

$$K = \frac{Q}{t_1 - t_2}$$

Consider, now, heat flowing along a bar (Fig. 1.3). At point X on the bar the temperature is t_1 and at point Y the temperature is t_2. The curve $t_1 - t_2$ indicates the way in which the temperature changes from X to Y along the bar. At points A, B and C, the temperature is t_a, t_b and t_c respectively. This is called the *temperature gradient* of the bar.

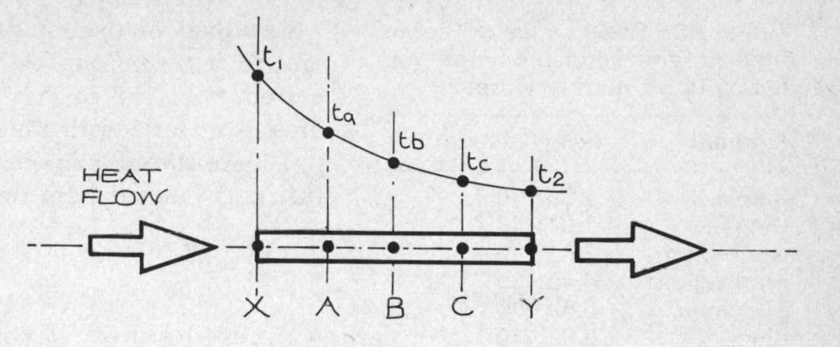

Fig. 1.3 Temperature gradient

Let Q = the heat flowing along the bar and discharged (watts)
a = the cross-sectional area of the bar (m^2)
L = the length of the bar (m)
t_1 = the temperature at X (°C)
t_2 = the temperature at Y (°C)
K = the conductivity of the material (W/m°C).

Then

$$K = \frac{Q}{a} \div \left(\frac{t_1 - t_2}{L}\right)$$

or

$$Q = \frac{K \times a(t_1 - t_2)}{L}$$

Thermal conductivities of various materials are shown in Table 1.2.

Example

A shaft 1 m long with a cross-sectional area of 0.1 m^2 is heated at one end to 100°C. The temperature at the opposite end is measured and is found to be 20°C. What is the rate at which the heat is conducted, given that K for the material is 236 W/m°C?

Solution

$$Q = \frac{236 \times 0.1 \times (100 - 20)}{1} = 1888 \text{ W} \quad (\textit{Ans})$$

1.3 Electrical Conductivity and Resistivity

The **electrical conductivity** of a substance is its ability to allow the passage of electric current along or through its length, and is proportional to its cross-sectional area. A good conductor has a high conductivity; a good resistor has low conductivity.

Table 1.3 Resistivities of some metallic and semiconductor materials (room temperature)

Element	*Resistivity* Ωm
Silver	1.59×10^{-8}
Copper	1.67×10^{-8}
Aluminium	2.65×10^{-8}
Magnesium	4.46×10^{-8}
Tungsten	5.50×10^{-8}
Zinc	5.92×10^{-8}
Nickel	6.89×10^{-8}
Iron	9.71×10^{-8}
Titanium	11.50×10^{-8}
Selenium	12.00×10^{-8}
Mercury	94.00×10^{-8}
Carbon	1.37×10^{-5}
Silicon	1.00×10^{-3}
Germanium	6.00
Boron	1.80×10^{4}

Table 1.4 Melting points

Element	*Melting Temperature* °C
Aluminium	657
Copper	1083
Magnesium	651
Tungsten	3370
Zinc	419
Nickel	1455
Iron	1535
Steel	1371
Eutectic soft solder	183

Resistivity on the other hand is the reciprocal of conductivity and, as its name implies, it is a measure of the opposition to current flow. It is inversely proportional to the cross-sectional area. A material with a high conductivity has a low resistivity.

The resistance R in ohms of a conductor at a given temperature is calculated from the formula

$$R = \rho \frac{L}{a}$$

where ρ = resistivity of conductor material (ohm-metre)
L = length of conductor (m)
a = cross-sectional area (m^2).

Table 1.3 gives the resistivity of various metallic elements and semiconductors at room temperature (20°C), arranged in order of increasing resistivity.

The resistance of pure metals increases with temperature. For example, if copper is raised from 20°C to 100°C, the resistance increases by about 30%.

Example

Calculate the resistance of a tie bar manufactured from iron and having a diameter of 14 mm and length 0.5 m.

Solution Resistivity of the cast iron is 9.71×10^{-8} Ωm.

$$\text{Cross-sectional area of the tie bar} = \frac{\pi \times 14^2}{4}$$

$$= 153.938 \text{ mm}^2$$

$$= 1.539 \times 10^{-4} \text{ m}$$

Therefore

$$\text{Resistance} = \frac{9.71 \times 10^{-8} \times 0.5}{1.539 \times 10^{-4}} = 3.155 \times 10^{-4}\ \Omega \quad (Ans)$$

1.4 Melting Point

The **melting point** of a solid substance is the temperature at which it changes to a liquid. This temperature is also the same as that at which solidification takes place. Different substances generally have different melting points. The melting point can be a critical design consideration if the product is to be used at elevated temperatures. Some typical melting temperatures of various substances are shown in Table 1.4.

1.5 Coefficient of Expansion

Most substances expand when the temperature is raised, and contract when cooled. This can sometimes be very useful for the designer, especially if it is necessary to assemble a shaft into a hub where an interference fit is required between the two components.

Table 1.5 Coefficient of linear expansion of some common engineering materials

Material	*Coefficient* per °C
Aluminium	25.5×10^{-6}
Brass	18.9×10^{-6}
Copper	16.7×10^{-6}
Cast iron	10.2×10^{-6}
Nickel steel (10% Ni)	13.0×10^{-6}
Mild steel	11.9×10^{-6}
Zinc	26.0×10^{-6}
Tin	21.4×10^{-6}

The **coefficient of linear expansion** of a substance can be defined, for a bar of unit length, as the increase in length when its temperature is raised by one degree.

Let α = the coefficient of linear expansion (per °C)
L = original length of component (m)
t = increase in temperature (°C)

Consider a metal bar of unit length.

Increase in a bar of unit length $= \alpha t$
Increase of bar length $L = L\alpha t$
Final length of bar $= L + (L\alpha t) = L(1 + \alpha t)$

Coefficients of linear expansion of some popular engineering materials, per degree centigrade, are given in Table 1.5.

As an alternative to heating a hub for assembly on a shaft, the shaft may be frozen in dry ice and allowed to cool sufficiently to enable it to pass through the bore of the hub. The above formulae holds good for this condition also.

Example

A steel bar having a total length of 30 cm is heated to a temperature of 168°C. The ambient temperature initially is 16°C. Find the new length of the bar if the coefficient of expansion is 13×10^{-6} per °C.

Solution

Increase in temperature $= (168 - 16) = 152$°C
Final length of bar $= 30(1 + 13 \times 10^{-6} \times 152)$
$= 30.059$ cm (*Ans*)

1.6 Magnetic Properties

Most materials are non-magnetic. Only iron, nickel, cobalt, ferric oxide, and alloys containing chromium, nickel, manganese or copper are magnetic to any extent.

High silicon-iron the most **permeable** material, i.e. it allows the passage of magnetic flux, and it is widely used in magnetic circuits, as in transformers.

The design specification would state any need to consider this aspect, but in most general mechanical design applications the effect of the material's magnetic properties can be ignored.

1.7 Tensile Strength

One of the most important characteristics of metals is their ability to deform plastically with little or no loss in strength. This phenomenon is known as the **ductility** of the material and varies according to the carbon content of the steel.

In the case of low-carbon annealed steel, the deformation of a component when subjected to a tensile load is

Elastic if, after the load is removed, the material returns to its original shape and dimensions, or
Plastic when the elastic range is exceeded and any further strain results in no significant increase in stress.

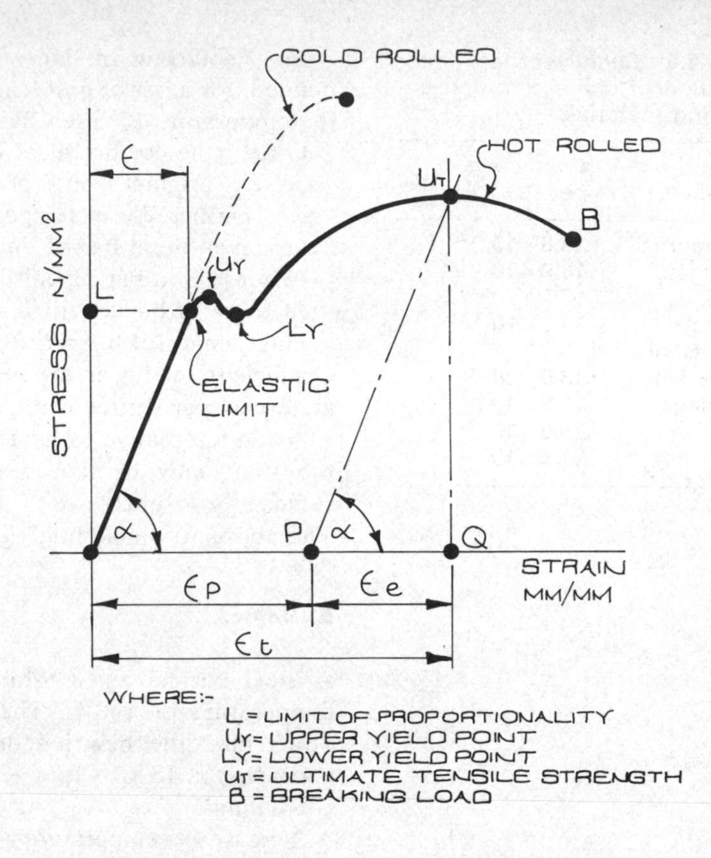

Fig. 1.4 Stress/strain graph

1.8 Tensile Tests

In testing a low carbon steel, a test piece is used to find the behaviour of the piece when subjected to a slowly applied tensile load. The test bar used is of uniform cross-section, usually circular. It is placed in a tensile testing machine where it is gripped in jaws at both ends and a pull is exerted axially. The load is increased slowly and a **stress/strain curve** is produced similar to that shown in Fig. 1.4.

If the material is stressed to point U_T, upon removal of the load, so that the stress returns to zero, then the plastic (permanent) strain is obtained, following downwards from point U_T to point P along a line approximately parallel to the line of proportionality.

At the base line, point Q, the total strain ϵ_t is composed of an elastic component ϵ_e, and a plastic component ϵ_p. The total area under the curve to B represents the total energy per unit volume required to break the specimen.

The dotted-line curve represents the result obtained when a cold rolled specimen is tested.

Figure 1.5 shows the state of necking, which results when the yield of the material is exceeded. This phenomenon increases beyond the ultimate tensile strength until finally the

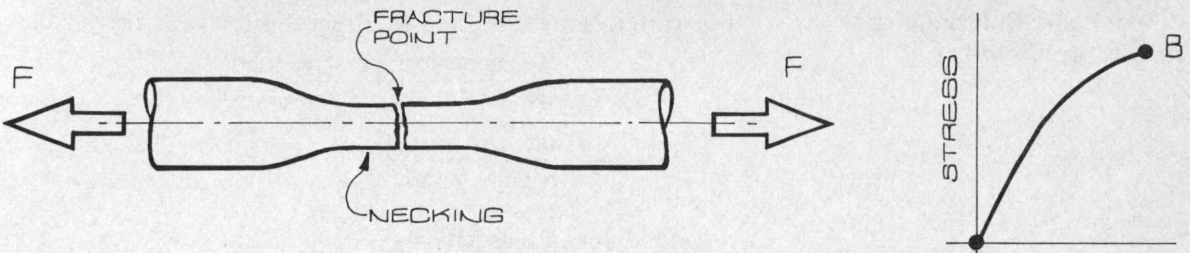

Fig. 1.5

Fig. 1.6 Stress/strain curve for cast iron

piece fractures, corresponding to point B in the stress/strain curve. The cross-sectional area reduces considerably between points U_T and B. Even with progressively lower load, extension increases until fracture occurs.

Figure 1.6 shows the stress/strain curve for cast iron. Fracture occurs without necking taking place, due to its more brittle nature.

1.9 Compressive Properties

A component undergoes deformation when subjected to a compressive force. Figure 1.7 shows how a low carbon annealed steel deforms when subjected to a compressive force.

Owing to friction f at the surfaces of contact, the metal does not deform uniformly but takes on a barrel shape. To avoid buckling, the test piece length must usually be less than twice the diameter.

For hot rolled annealed mild steel, the stress/strain curve is shown in Fig. 1.8.

The stress/strain curve for compression is roughly similar to the tensile one, and stresses may be calculated in the same way for both load conditions because the modulus of elasticity is about the same for both conditions.

A well-defined yield point occurs, after which stress continues to rise with increasing strain. No maximum load or stress is reached.

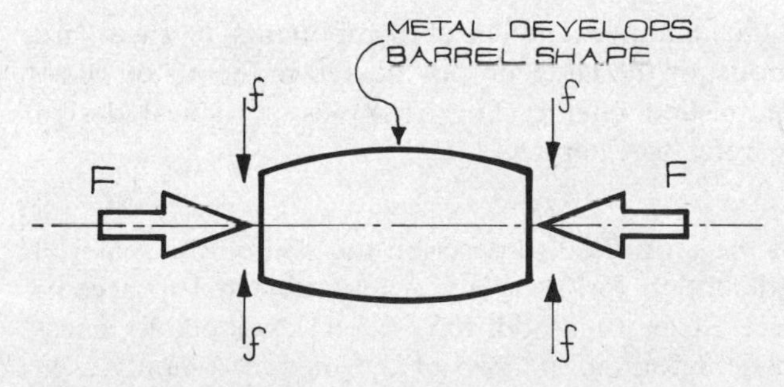

Fig. 1.7 Deformation under compression

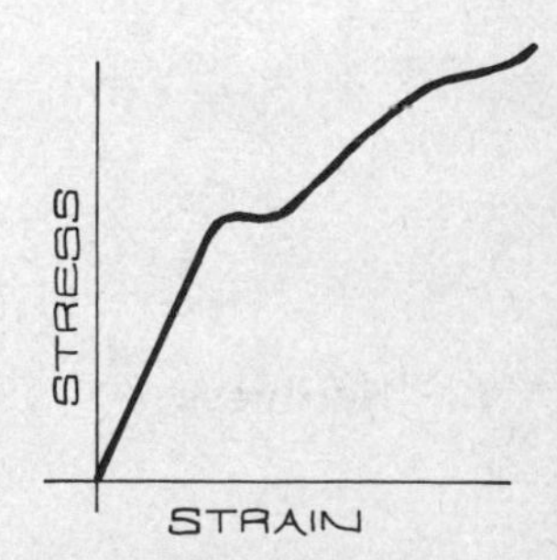

Fig. 1.8 Stress/strain curve for hot rolled mild steel

1.10 Basic Relationships in Simple Stress

For components subjected to direct tension or compression:

$$\text{Stress} = \frac{\text{Tensile or compressive load}}{\text{Cross-sectional area under stress}}$$

$$\text{Strain} = \frac{\text{Change in length}}{\text{Original length}}$$

$$\text{Modulus of elasticity} = \frac{\text{Stress}}{\text{Strain}}$$

Example

A tie bar on a vertical pressing machine is 2 m long and 38 mm diameter. What are the stress and the extension under a load of 100 000 N?

The modulus of elasticity is 208 300 N/mm^2.

Solution

$$\text{Cross-sectional area } A = \frac{\pi \times 38^2}{4} = 1134.12 \text{ mm}^2$$

$$\text{Stress} = \frac{100\,000}{1134.12} = 88.17 \text{ N/mm}^2 \quad (\textit{Ans})$$

Original length = 2 m = 2000 mm

Modulus of elasticity $E = 208\,300$ N/mm^2

$$\text{Therefore} \quad \text{Strain} = \frac{88.17}{208\,300} = 4.233 \times 10^{-4}$$

$$\text{Change in length} = \text{Extension} = 4.233 \times 10^{-4} \times 2000$$

$$= 0.85 \text{ mm} \quad (\textit{Ans})$$

1.11 Shear Strength

When a load q is applied which acts in a straight line, together with an equal and opposite load of the same magnitude as q, the two forces being distance Δy apart, then the body is said to be in **shear**. Fig. 1.9 illustrates this load condition.

$$\text{Average shear stress } q_A = \frac{q}{A}$$

The most common examples of components in shear are rivets and bolts, or the gudgeon pin located in the piston of an internal combustion engine. For examples of typical design applications refer to Chapter 2.

1.12 Hardness

Hardness is the term used to describe the ability of a material to resist indentation to its surface when subjected to a compressive force. Several methods have been developed to measure the plastic deformation. Two of the most commonly used ones are described below.

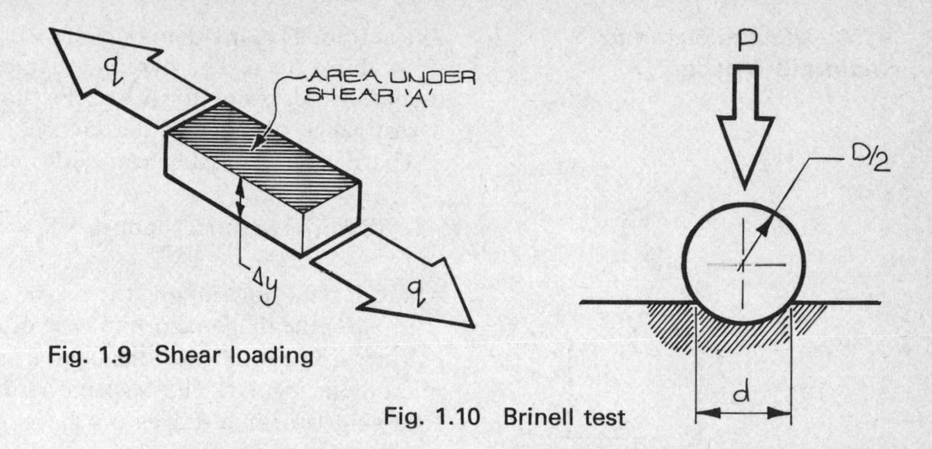

Fig. 1.9 Shear loading

Fig. 1.10 Brinell test

1.13 Brinell Test

In this test, a hardened steel ball is pressed into the surface of the test piece. Figure 1.10 shows the ball indenting a surface and the notation used to establish the hardness number.

The average diameter of the impression is measured and

$$\text{Brinell hardness number} = \frac{\text{Load in kilogrammes}}{\text{Curved surface area (mm}^2\text{)}}$$

where the curved surface area is approximately

$$A \approx \tfrac{1}{4}d^2\left(1+\frac{d^2}{4D^2}\right)$$

and $\text{BHN} = \dfrac{P}{A}$

The standard load for testing steel is 3000 kg with a 10 mm ball. For soft materials this load is too high, and a smaller ball and load are used.

For other materials P should be varied by applying the relationship

$$\frac{P}{D^2} = \text{constant}$$

where the constant is

Steel and cast iron = 30
Copper alloys = 10
Aluminium alloys = 10
Copper and aluminium = 5

Example For a copper alloy using a 2 mm diameter ball, find the load required

$$\frac{P}{2^2} = 10 \qquad \text{therefore} \quad P = 40\ \text{kg}$$

1.14 Vickers Pyramid Diamond Method

An extremely hard diamond-shaped indentor is used to impress the surface. The pyramid diamond is a 136° square-base indentor and, owing to its extreme hardness, can be used over a vast range of material hardness.

The hardness number can be found from the formula

$$\text{Vickers Pyramid Number VPN} = \frac{2 \times P \times \sin 68°}{d^2}$$

where P = the load applied

d = the diagonal of the base of the pyramid (Fig. 1.11).

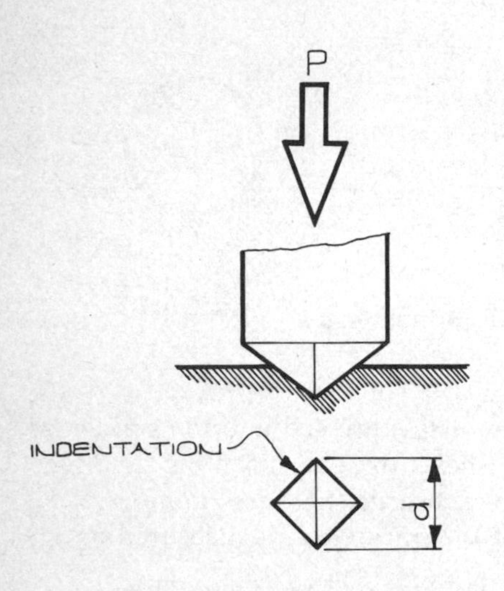

Fig. 1.11 Vickers pyramid diamond

Measurement of the diagonal is made with the aid of a micrometer ocular. The Vickers hardness number is found by focussing two knife-edges on the edge of the diagonals. The micrometer setting is read from a digit counter on the measuring machine and is converted with the aid of tables to give the correct VPN.

The Brinell test has some defects and tends to give a hardness reading less than the hardness of the material undergoing test. Therefore the Vickers test is preferred for the harder materials.

Figure 1.12 summarises the hardness tests.

1.15 Impact Test

Static tests are unsuitable for determining the ability of a material to resist shock loads. Methods have been developed to establish shock resistance by using a notched or unnotched specimen. They are carried out by placing a specimen in an anvil and allowing a hammer to swing about its fulcrum and strike the specimen. The hammers swing then breaks the specimen and continues some distance on the other side.

In the IZOD test, the potential energy of the fall is set at 16.6 kg and a free pointer records the energy absorbed by the specimen to break it.

In the CHARPY test, the specimen is held at both ends and struck by a wedge-shaped hammer at the centre.

These tests are invaluable in finding the temper brittleness in heat-tested nickel-chrome steels and reveal the resistance of the materials to stress concentrations in a component.

The more brittle the material, the less energy required to break the specimen.

Figure 1.13 shows the IZOD and CHARPY impact testing machines.

1.16 Design Example

Having established the design form, the functional requirements and the operating environment, the designer must ensure that the material selected possesses the right mechanical, physical and chemical characteristics to meet the application

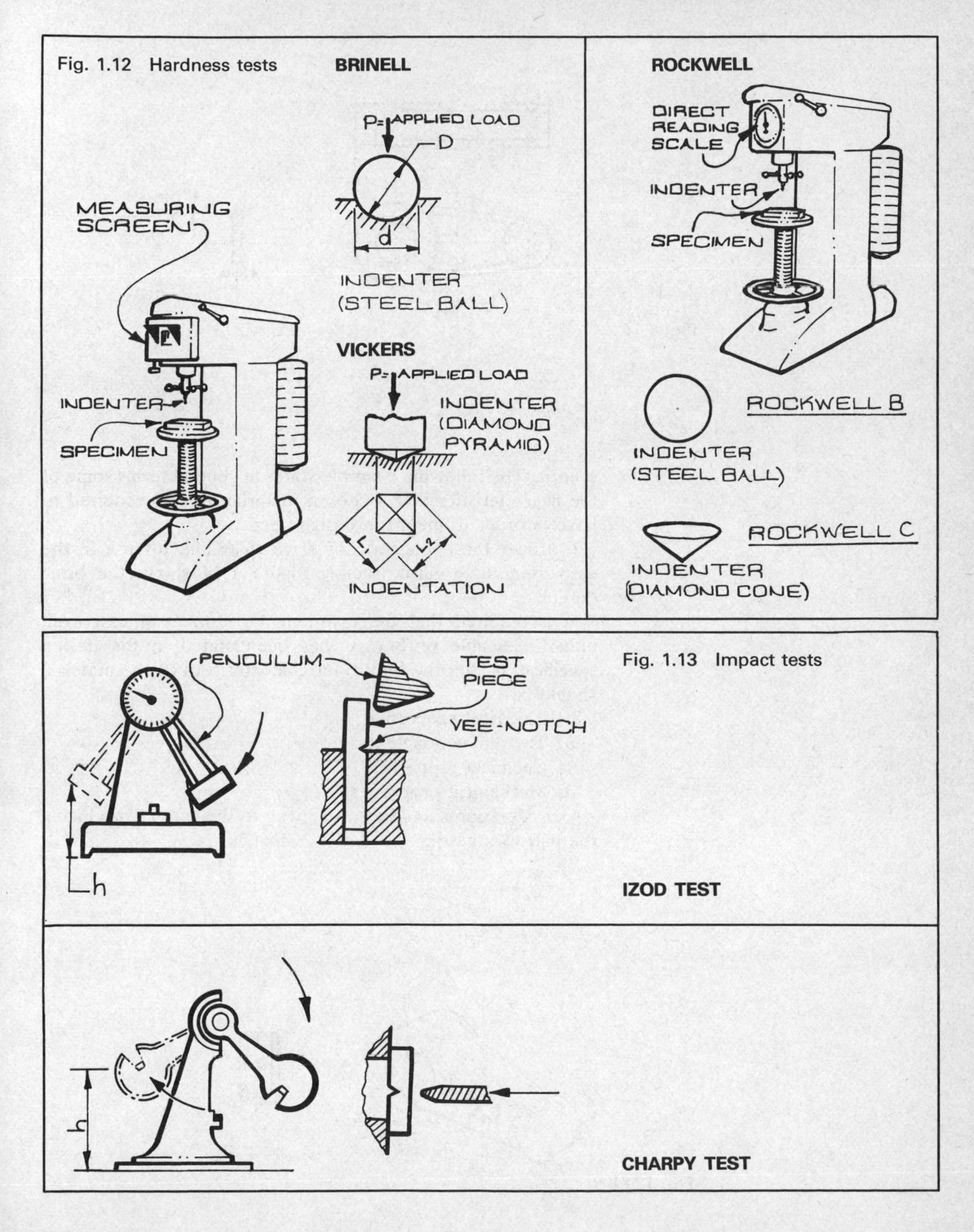

Fig. 1.12 Hardness tests

Fig. 1.13 Impact tests

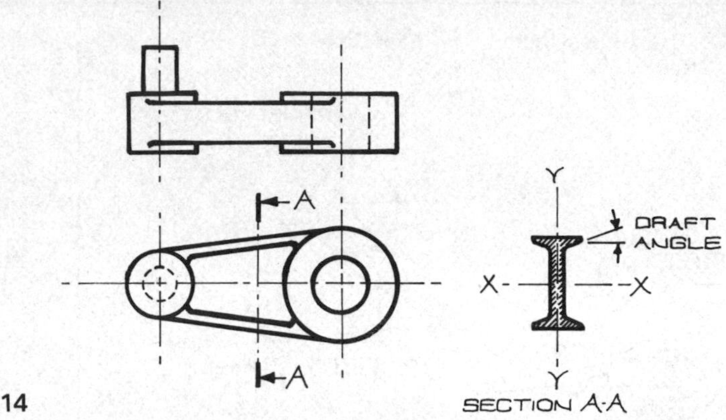

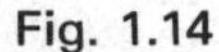
Fig. 1.14

criteria. The following example states in general terms some of the characteristics that a chosen material may be required to have in order to meet the design need.

Consider the application of drive shaft link for use in the suspension of a farm tractor. Figure 1.14 shows the basic design.

It is required that the component be made of cast iron, either malleable or SG. It has been stated in the design specification that the important characteristics of the material should be

a) Electrical properties
b) Thermal conductivity
c) Chemical properties
d) Mechanical properties.

Consideration should also be given to the ease of producing the part as a casting.

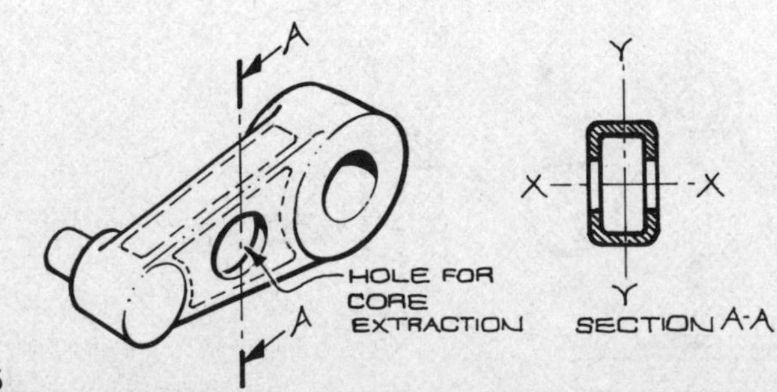

Fig. 1.15

PRODUCTION By designing the shape as an I-section, the casting is simplified since the necessary cores can be made so that they will be separable from the component after casting, without the need to provide special core-removal techniques. If the cross-section is as shown in Fig. 1.15, then special core holes in the body of the casting would be needed in order to remove the core.

MECHANICAL PROPERTIES The modulus of rupture about the X axis would be about the same for both sections. The modulus of rupture about the Y axis for Fig. 1.14, however, would be approximately twice that for the section shown in Fig. 1.15. Therefore the I-section is not only easier to produce but gives better bending resistance. If the component was put in torsion at the cross-section indicated, the I-section would give only approximately one tenth the resistance that the box section gives.

ELECTRICAL PROPERTIES The electrical resistivity of malleable iron is lower than SG iron due to the lower carbon and silicon content. A typical value is taken as 0.32 $\mu\Omega$m.

THERMAL CONDUCTIVITY The thermal conductivity properties of irons containing a flake structure tend to be higher than those containing a nodular (SG) structure. Malleable irons have a thermal conductivity somewhere in between the modular and flake graphite irons.

At 100°C the value is approx. 49 W/m°K for ferritic blackheart malleable iron, and for pearlitic malleable iron it is approx. 39.5 W/m°K.

COEFFICIENT OF THERMAL EXPANSION Malleable irons have coefficients of thermal expansion similar to flake graphite irons. Between 20° and 200°C, the value is 11×10^{-6} per °K.

CHEMICAL PROPERTIES A structural iron is required to be resistant to water. Cast iron in itself is not resistant to rusting where moisture is present. Suitable surface treatments are available for protective purposes, and it is the designer's responsibility to take the necessary steps to ensure the durability of the component in service. Electroplating and painting are typical of surface protective treatments.

Problems

1.1 What is the mass of a circular bar made from copper, when the bar is 3 cm diameter and 5 m long, and the density of copper is 8957 kg/m^3?

1.2 If the mass of an aluminium tube is 10 000 kg, and it is 20 cm outside diameter and 15 cm inside diameter, how long is the tube if its density is 2568 kg/m^3?

1.3 A bar has a length of 2.5 m and is at an ambient temperature of 20°C. If it is heated to 2000°C, what is the increase in length if the coefficient of linear expansion of the material is 11.9×10^{-6}?

1.4 A metal object of dimension 75 mm × 32 mm × 16 mm is heated from 20°C to 72°C. If the new length is 75.06 mm, what is the coefficient of linear expansion?

1.5 What is the heat conducted by a copper shaft, 10 cm diameter and 2 m long, if it is heated from 30°C to 100°C? The material thermal conductivity is 403 W/m°C.

1.6 Define the relationship between stress, strain and the cross-sectional area.

1.7 A brass tube 50 mm outside diameter, 43 mm inside diameter, and 300 mm long is compressed between two end plates by a load of 20 000 N. The reduction in length is measured as 0.15 mm. Calculate the modulus of elasticity.

1.8 What material is suitable for a bar having a diameter of 28 mm, length 1 m, and electrical resistance $4.3 \times 10^{-5}\ \Omega$. (Refer to Table 1.3.)

1.9 Calculate the resistance of a bar of length 2.6 m having a cross-sectional area of 0.015 m and a resistivity of 9.71×10^{-8}.

1.10 Calculate the strain in a bar of length 2.3 m when extended to 2.310 m.

2 Strength and Selection of Materials

Engineering materials may be broadly classified under the headings of ferrous metals, non-ferrous metals, plastics.

2.1 Ferrous Metals

A ferrous metal is one which contains iron. To the engineer this means either a steel or a cast iron. In both cases the other essential ingredient is carbon, but it is the quantity and form of carbon which govern the type of material. Cast iron usually contains between 2.5 and 4% carbon. This is more than for steels which contain up to 1.7% carbon and require much more critical control of percentage content.

2.2 Grey Cast Iron

This is the most widely used of all cast irons and is thus often referred to merely as **cast iron**. Its carbon takes the form of pointed graphite flakes in an iron background. This discontinuous structure is largely responsible for the general properties of the metal. These properties may be listed as follows:

1 Excellent fluidity in casting, enabling intricate shapes to be cast quickly and cheaply, eg. cylinder heads, motor frames, manifolds.

2 Weakness in tension, due to the graphite structure. The points on the graphite flakes cause "stress-raisers" which result in a tensile strength lower than steels.

3 High compressive strength and rigidity, making it the essential material for large machine-tool beds.

4 Low coefficient of friction. The graphite content gives grey cast iron self-lubricating properties, useful for application such as bearing blocks and piston rings.

5 Good machining properties, due to the brittleness of the flake structure.

6 Vibration damping properties, giving advantage in applications such as gearbox housings.

Grey cast iron is supplied almost entirely in the form of sand castings. It is low in ductility and cannot be rolled or drawn.

2.3 Malleable Cast Irons

These are produced by heat-treating a material called white cast iron (an extremely hard iron-carbide structure usually produced by rapid cooling from a furnace). The chief objective in producing **malleable irons** is to obtain a material with the castability of grey cast iron, but with higher tensile strength and a certain amount of ductility. The two most common types of malleable iron are

1 Whiteheart malleable iron. This is produced by a decarburizing heat-treatment process. In thin sections the carbon is almost entirely extracted, leaving almost pure iron. This restricts whiteheart malleable to thin components such as pipe fittings, and hinges. Thicker sections should be avoided due to a hard discontinuous core which often forms inside the decarburized structure.

2 Blackheart malleable iron. In this process, white cast iron is heat-treated to 900°C. This results in a graphite structure in the form of "rosettes" which have less severe stress-raisers than in those of the graphite flakes in grey cast iron. Blackheart malleable iron is widely used for smaller components in the automobile industry.

2.4 Spheroidal Graphite Cast Iron

This is also referred to as **SG Iron** or Nodular Iron. Its structure takes the form of graphite spheres produced by chemical reaction when magnesium is added to the molten metal in the ladle prior to casting. Unlike malleable irons, there is no limit to the size of an SG iron casting.

This material is expensive but has excellent mechanical properties, combining high tensile strength, high compressive strength, toughness and ductility with excellent castability and machinability. In applications such as crankshafts, connecting rods, and gears, SG iron is often replacing high tensile forged steel.

2.5 Steels

The versatility of steel as an engineering material is due to its wide range of good mechanical properties and forming processes. Steels generally combine most of the following properties:

a) High tensile strength
b) High compressive and shear strength
c) Good ductility and malleability
d) Toughness and hardness
e) Moderately high thermal and electrical conductivity

f) High melting point
g) Good machinability
h) Ease of heat treatment.

Unlike grey cast iron, steels are easily rolled, drawn, and pressed into shape. Typical forms of supply are hot-rolled, cold-rolled and cold-drawn sections, forgings, and to a lesser extent, extrusions and castings.

2.6 British Standard En Steels

BS 970:1955 classified types of steel under **En numbers**. Under the appropriate En number was found the composition of the steel, its mechanical properties, and its reaction to heat-treatment, welding and machining. This classification system has now been replaced by that shown in BS 970:1973. However, at the time of publication, the En system is still widely used in industry. Therefore, the steels mentioned in the following paragraphs will be given under both classifications. However, in many cases the alternatives are not exact equivalents but represent close alternative steels in respect of alloying elements and mechanical properties.

2.7 Plain Carbon Steels

A steel is usually classed as a **plain carbon steel** if constituents other than iron and carbon are not in excess of 1.5%. Plain carbon steels may be broadly grouped as follows:

1 Low carbon steels (up to 0.25% carbon) These are also called **mild steels** and generally have moderate strength and toughness, good ductility, and are easily machined and welded.
2 Medium carbon steels (0.25–0.65% carbon) These are characterised by their high strength and toughness, and their good response to heat treatment.
3 High carbon steels (0.65%–1.7% carbon) These are very hard, especially when heat-treated.

The following are some common plain carbon En steels:

(220M07) En1A Cold-formed mild steel used for low duty nuts, bolts, studs, etc.

(070M20) En3 Hot rolled, general purpose mild steel. This is one of the most widely used of steels for mechanical and structural applications.

(080M40) En8 Medium carbon steel which has high strength and toughness, especially when heat-treated. Used extensively for keys, general forgings, crankshafts, connecting rods, etc.

(045M10) En32A	This is a common case-hardening steel used mainly for wear-resistant applications such as camshafts, tappets, gears, gudgeon pins, etc.
(070A72) En42	High carbon steel used chiefly for leaf springs and coil springs.

2.8 Alloy Steels

An **alloy steel** is one in which an element other than iron and carbon has been deliberately added in order to improve particular properties of the steel. Some common alloy steels are as follows:

Tungsten steels (high speed steel) These contain up to 20% tungsten and retain hardness and toughness properties at elevated temperatures. They are used mainly for high-speed cutting tools such as drills, saw blades, lathe tools, etc.

Chromium steels (e.g. En11 (526M60), En31 (535A99)) These contain up to 5% chromium and are renowned for their hardness and wear resistance. Typical applications are ball-bearings and gears.

Nickel-chromium steels (e.g. En24 (817M40)) These have excellent mechanical properties all-round. They have very high strength, toughness and hardness, good ductility and may be heat-treated. They are used for heavy duty applications such as highly stressed gears, crankshafts and aircraft parts.

Stainless steels These contain chromium in excess of 11%, which gives the material anti-corrosive properties. A vast range of stainless steels exist of varying chromium content and mechanical properties. Common uses of stainless steel include cutlery, sink tops, tubing, sheeting and surgical instruments.

2.9 Non-ferrous Metals

With a few notable exceptions, the common non-ferrous metals used in engineering tend to be soft and ductile and comparatively low in strength, toughness, and hardness.

2.10 Aluminium

This is the most plentiful metal on the earth's surface and in its pure form has the properties of very low density, good ductility, low strength, good corrosion resistance and good electrical conductivity. It is usually supplied in extruded, drawn, or cast form. Common applications are electric cables and general domestic use.

2.11 Aluminium Alloys

The usual objective here is to produce a material with the low density of pure aluminium and the strength and toughness of ferrous metals. The many types of aluminium alloy have been classified under BS 1470–77, 1490. These British Standard specifications broadly divide the alloys into

a) Wrought heat-treatable.
b) Wrought non-heat-treatable.
c) Cast heat-treatable.
d) Cast non-heat-treatable.

Cast aluminium alloy specifications are always preceded by the letters LM.

Some common aluminium alloys are as follows:

Duralumin This is a wrought aluminium alloy which contains up to 4% copper and can be heat-treated. It is strong and tough with a density almost as low as pure aluminium. Its chief application is for aircraft parts.

LM6 (Alpax) Cast non-heat-treatable. This the most widely used aluminium alloy with excellent casting properties, low density and moderate strength. LM6 is supplied in sand-cast or die-cast form for general engineering applications such as sumps, gearboxes, motor frames, carburettors, etc.

LM14 (Y alloy) Cast heat-treatable aluminium alloy for high-strength applications. LM14 is used chiefly for components such as pistons and cylinder heads which require strength, lightness, and good thermal conductivity to aid quick heat dissipation.

2.12 Copper

High-purity copper is noted for its excellent electrical conductivity, good thermal conductivity, good ductility, softness and corrosion resistance, and low mechanical strength. Even very small amounts of impurity, however, will seriously affect these properties. Although copper is a superior electrical conductor to aluminium, copper tends to be more expensive. Due to its good ductility, copper is usually supplied in cold-drawn form, eg. electrical wire, pipe for plumbing and general pipe fittings.

2.13 Copper Alloys

These usually come under the headings of brass (copper and zinc) and bronze (copper and tin).

Brasses There are a wide range of commercial brasses with varying properties according to the zinc content. Generally they are classified into two metallurgical groups: α-brasses containing up to 37% zinc; and $\alpha+\beta$ brasses containing more than 37% zinc. α-brasses are ductile and used for cold-working. A typical α-brass is "cartridge brass" containing 30% zinc. Cartridge brass is used for deep-drawn components, which require dimensional accuracy, corrosion resistance, and superior surface finish.

The most common $\alpha+\beta$ brass is **muntz metal** containing 40% zinc. This a hot-working metal which has good strength, hardness and toughness, whilst retaining moderate electrical conductivity and resistance to corrosion. Muntz metal is supplied as forgings, castings and extensions for such applications as pipe joints, electrical connections, nuts and screws, pump bodies, etc.

Phosphor bronze This contains copper, up to 15% tin and up to 1% phosphorus. The resulting alloy is hard, rigid, and has a low coefficient of friction. Thus phosphor bronze is used extensively for plain friction bearings.

2.14 Other Metals

Lead This is a soft, weak metal of high density. Its chief uses are in battery cells and as an essential ingredient of soft solder.

Nickel From the mechanical point of view, the chief application of nickel is that of an alloying element for other metals, particularly steels.

Tin This is a soft, weak metal with a very low melting point and good corrosion resistance. Its chief uses are as as an alloying element in bronze and soft solder, and as a protective coating for steel.

Zinc This is a weak metal used mainly as an alloying element of brass and as a protective coating for steel in the form of galvanizing or sherardizing (see Chapter 4).

When zinc is alloyed with small amounts of copper and aluminium, an excellent die-casting material is produced. Zinc die-castings have moderate strength and toughness. BS 1004 classified these metals into two grades known as Alloy A and Alloy B, with slightly differing metallic composition. Alloy B is stronger and harder, but is more affected by heat. Both alloys are subject to prolonged deformation after manufacture, but this can be overcome with suitable heat-treatment. Zinc alloy die-castings are extensively used in small parts for domestic items, and small car components such as carburettor parts and door hinges.

2.15 Plastics

A plastics material is defined as a chemically-produced organic (carbon-base) material which has been moulded to shape during its manufacture. In considering the possible use of a plastic, the engineering designer should consider the following general properties of plastics in comparison with metals:

a) Usually lower strength and toughness.
b) Lower density.
c) Usually lower melting point properties, more seriously affected by high temperature.
d) Electrical insulator and poor conductor of heat.
e) Non-corrosive.
f) Wide range of colours; can be transparent.

Plastics are normally classified under two headings: thermoplastics and thermosetts. A **thermoplastic** will liquefy under the action of heat, solidify when cooled, and will liquefy again if more heat is applied. Thus a thermoplastic may be moulded and re-moulded at will. A **thermosett** is liquid only during manufacture. During solidification, a chemical reaction takes place preventing any subsequent melting. If heat is re-applied, a thermosett will not melt, although mechanical properties will be affected. Thermosetts tend to be hard, brittle and more rigid than the flexible thermoplastics.

2.16 Some Common Thermoplastics

Nylon This is the most widely used plastic in mechanical engineering. It is tough, rigid, wear-resistant and has a higher melting point than most other plastics. Unfortunately its high coefficient expansion causes problems with dimensional accuracy at elevated temperatures. Also, tensile strength rapidly falls with temperature rise. Of the many commercial nylons available, the most common is Nylon Grade 66 which is used extensively for low duty gears, bearings, couplings and washers.

PVC (Poly-vinyl-chloride) Tough, flexible and non-inflammable. PVC is used mainly for electric wire insulation and guttering.

PTFE (Poly-tetra-fluoro-ethylene) This has a low coefficient of friction and is used mainly for bearings and seals.

2.17 Some Common Thermosetts

Polyester This often bonded with fibre glass and used for car bodies and protective helmets

Phenolic laminates (e.g. "Tufnol") Cotton fibre is impregnated with phenolic resin and several layers are then pressed together, producing a strong laminated sheet. This material has wide use for items requiring mechanical strength and electrical resistance. It is often used for smooth-running gears especially in electrical equipment.

Ebonite Hard, brittle plastic, commonly used for general electrical hardware and contrast knobs.

2.18 Design Dimensions and Factors of Safety

Components carrying mechanical loads will be subjected to various types of stress conditions. The designer must therefore locate the weakest point of the component, recognise the particular type of stress, and calculate the minimum allowable dimension.

Tensile, compressive and shear strength values are determined under ideal static conditions. In practise however a component is subjected to numerous types of shock loading.

Thus for a particular application, the designer must apply a **factor of safety** to the strength figure quoted for a material. The value of this factor will vary according to the severity of the shock loading, and when applied gives the allowable stress for the application. Therefore

$$\text{Allowable stress} = \frac{\text{Material strength}}{\text{Factor of safety}}$$

Some of the more common design calculations for components are now given.

2.19 Bolts, Screws and Pins

These could fail due to tension, shear and, to a lesser extent, compression. In each case it is essential to recognise the area subjected to the stress.

Example 1

A pin made of En3 steel is subjected to a tensile load of 10 000 N with a required safety factor of 6. Calculate the minimum diameter of the pin.

Solution Tensile strength of En3 = 460 N/mm^2 (see Chart 9)

$$\text{Allowable stress} = \frac{460}{6} = 76.7\ \text{N/mm}^2$$

$$\text{Stress} = \frac{\text{Force}}{\text{Area}} \qquad \text{Area} = \frac{\text{Force}}{\text{Stress}}$$

$$\text{Minimum area} = \frac{10\,000}{76.7} = 130.4\ \text{mm}^2$$

$$\text{Area } A = \pi r^2$$

$$r = \sqrt{\frac{A}{\pi}} = \sqrt{\frac{130.4}{\pi}} = 6.44\ \text{mm}$$

Minimum diameter = 12.88 mm (*Ans*)

The pin would be designed to no less than 13 mm diameter.

Example 2

A number of M12 bolts made of En1A are subjected to a tensile load of 25 000 N with a required safety factor of 8. Find the minimum number of bolts required.

Solution As is common with bolts and screws, the designer

is constrained to use the standard sizes of bolts and thus is concerned with the minimum number required. However this number will undoubtedly affect other design dimensions. It should be realized that the area under tension is at the *minor diameter* of the thread.

$$\text{Tensile strength of En1A} = 420\ \text{N/mm}^2 \text{ (see Chart 9)}$$

$$\text{Allowable stress} = \frac{420}{8} = 52.5\ \text{N/mm}^2$$

Area under tension is given by using the minor diameter of M12 bolt.

$$\text{Minor diameter} = 9.853\ \text{mm} \quad \text{(see Chart 11)}$$

$$\text{Area of bolt} = \pi r^2 = \pi(4.926^2) = 76.23\ \text{mm}^2$$

$$\text{Stress} = \frac{\text{Force}}{\text{Area}} \qquad \text{Area} = \frac{\text{Force}}{\text{Stress}}$$

$$\text{Minimum total area} = \frac{25\,000}{52.5} = 476.2\ \text{mm}^2$$

$$\text{Number of bolts} = \frac{476.2}{76.23} = 6.247$$

Therefore, 7 bolts are required (*Ans*)

Example 3

The 4000 N force shown in Fig. 2.1 is applied to the M10 setscrew. If the required safety factor is 3, decide on a suitable material for the setscrew.

Solution Type of loading is shear.

Area under stress is given by minor diameter of M10.

$$\text{Minor diameter} = 8.16\ \text{mm (see Chart 11)}$$

$$\text{Area of screw} = \pi r^2 = \pi(4.08)^2 = 52.3\ \text{mm}^2$$

$$\text{Shear stress} = \frac{4000}{52.3} = 76.48\ \text{N/mm}^2$$

Try En1A as the material.

$$\text{Shear strength} = 250\ \text{N/mm}^2 \text{ (see Chart 9)}$$

$$\text{Allowable stress} = \frac{250}{3} = 83.3\ \text{N/mm}^2$$

Therefore, En1A is adequate and is the material chosen for the setscrew.

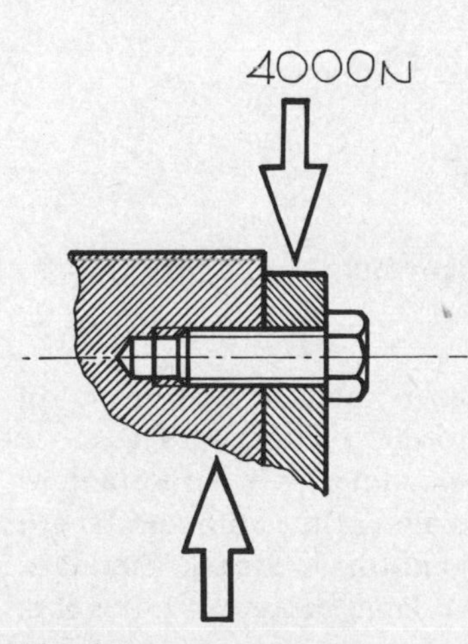

Fig. 2.1

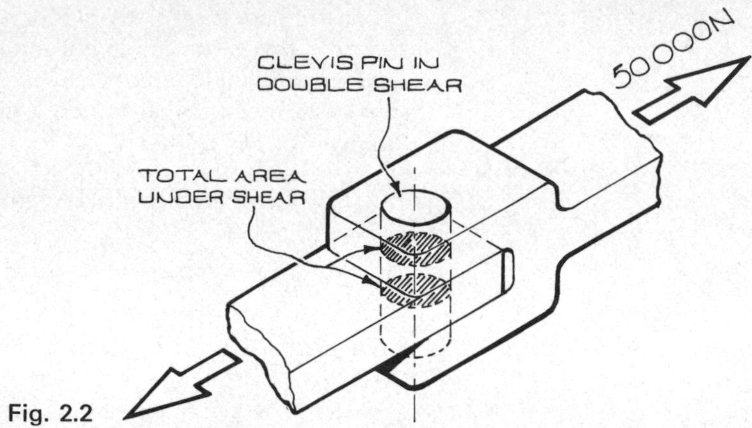

Fig. 2.2

Example 4

A force of 50 000 N is applied to the clevis pin shown in Fig. 2.2 with a required safety factor of 4. If the pin is to be made of En3, determine its minimum allowable diameter.

Solution This pin is subjected to a condition known as *double shear*, i.e. the shear load is taken by two areas. This means that the available area is doubled and thus the stress is halved. For equal diameters, a condition of double shear is obviously mechanically sounder than one of single shear.

Shear strength of En3 = 270 N/mm² (see Chart 9)

$$\text{Allowable stress} = \frac{270}{4} = 67.5\ \text{N/mm}^2$$

$$\text{Required total area} = \frac{\text{Force}}{\text{Stress}} = \frac{50\,000}{67.5} = 740.7\ \text{mm}^2$$

$$\text{Area at one point } A = \frac{740.7}{2} = 370.4\ \text{mm}^2$$

$$r = \sqrt{\frac{A}{\pi}} = \sqrt{\frac{370.4}{\pi}} = 10.86\ \text{mm}$$

Therefore, minimum allowable diameter = 21.72 mm (*Ans*)

2.20 Keys

Shaft keys support items such as gears, pulleys, and levers on shafts which are subjected to a torque condition. Therefore they invariably fail due to *shear* loading. It is important to realize that the area under shear is along the width and length of the key. The designer usually conforms to British Standard recommendations on the width and depth of key and thus it is the length which must be calculated.

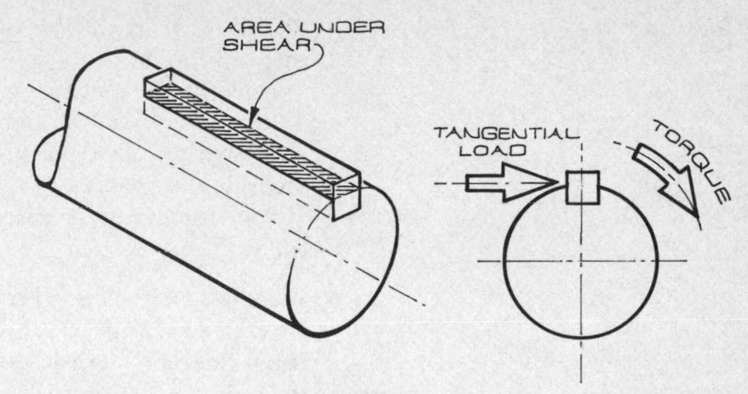

Fig. 2.3

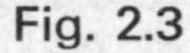

Example

A 40 mm diameter shaft (Fig. 2.3) is subjected to a tangential load of 20 000 N around its circumference. If the required safety factor is 6, decide on a suitable material and overall dimensions for the key.

Solution Chosen material is En8 (see Chart 9).

Shear strength of En8 = 370 N/mm^2 (see Chart 9)

$$\text{Allowable stress} = \frac{370}{6} = 61.2\ \text{N/mm}^2$$

For 40 mm diameter shaft:

Width of key = 12 mm, Depth of key = 8 mm (see Chart 12)

$$\text{Allowable area} = \frac{\text{Force}}{\text{Stress}} = \frac{20\,000}{61.2} = 326.8\ \text{mm}^2$$

$$\text{Width} \times \text{length} = 326.8\ \text{mm}^2$$

$$12 \times \text{length} = 326.8\ \text{mm}^2$$

$$\text{Minimum length} = \frac{326.8}{12} = 27.2\ \text{mm}$$

Chosen key specification: Material En8; Width 12 mm; Depth 8 mm; Length 28 mm (*Ans*).

2.21 Structural Steel Sections

These may be subjected to a varied number of complex loading conditions. However, in most cases they will fail either due to bending (a combination of tension and compression) or torsion (a type of shear loading due to the twisting effect of a torque). Although bending and torsion calculations will not be covered here, it will be of interest to note that the shape of cross-section has considerable effect on the stress condition.

2.22 Shafts

Shafts are nearly always of circular cross-section and could fail due to bending or torsion. The stress condition will be greatly affected by the machining characteristics of the finished items. Sharp corners and abrupt changes in diameter tend to cause stress-raisers, giving a much weaker design and greatly increasing the chances of failure.

Some of the more common features of shaft design are as follows.

Chamfers (Fig. 2.4) These remove sharp edges from external covers and aid the mating of male and female parts.

Radii (Fig. 2.5) All internal corners will have a radius due to the radius of the cutting tool. The corner radius will often have to lie within a tolerance range. Where the corner is a seating for a female component it is essential that the shaft radius is less than female chamfers. Similarly on a bearing seating, the shaft radius should be less than the bearing radius. However, a corner radius which is minute will create a stress-raiser weakness in the shaft and is thus to be avoided.

Undercuts (Fig. 2.6) These are often used on threaded ends to ensure complete clamping of female components. Excessively deep undercuts will cause stress-raiser weaknesses.

Severe changes in diameter (Fig. 2.7) These can cause stress-raisers and thus should be made gradual.

Problems

2.1 Thoroughly read the foregoing chapter concerning the description of common engineering materials and then using brief notes complete the blank spaces in the column headed "General Description and Uses" of the materials properties Charts 8, 9 and 10 (pages 146–148).

Note All following questions require reference to the completed materials Charts 8, 9, 10, and also to the various British Standards extracts at the back of this book, i.e. Charts 11 and 12.

2.2 *a*) A 14 mm dia. steel bar is subjected to a tensile load of 9240 N; calculate the stress.
b) Choose a suitable steel if the required SF is 6.

2.3 A structure suspended by En3 tie-bars is equivalent to a load of 100 000 N. The tie-bars each have a diameter of 28 mm. If the SF is 10, calculate the minimum number of tie-bars.

2.4 In a cylinder of bore 70 mm the pressure is 4 N/mm^2. The cover plate is secured by M10 studs with a required SF of 12. Determine the minimum number of studs using *a*) En1A, *b*) En3.

2.5 What diameter of En8 shaft will withstand a tensile load of 50 000 N if the required SF is 8?.

2.6 A 70 mm dia. shaft is subjected to a tangential load of 80 kN. With a SF of 5, design a suitable feather key for the shaft.

2.7 The maximum explosion pressure in the cylinder of a combustion engine is 0.8 N/mm^2. The piston has a diameter of 100 mm. Determine the minimum diameter of gudgeon pin, using a suitable steel, with a required SF of 16.

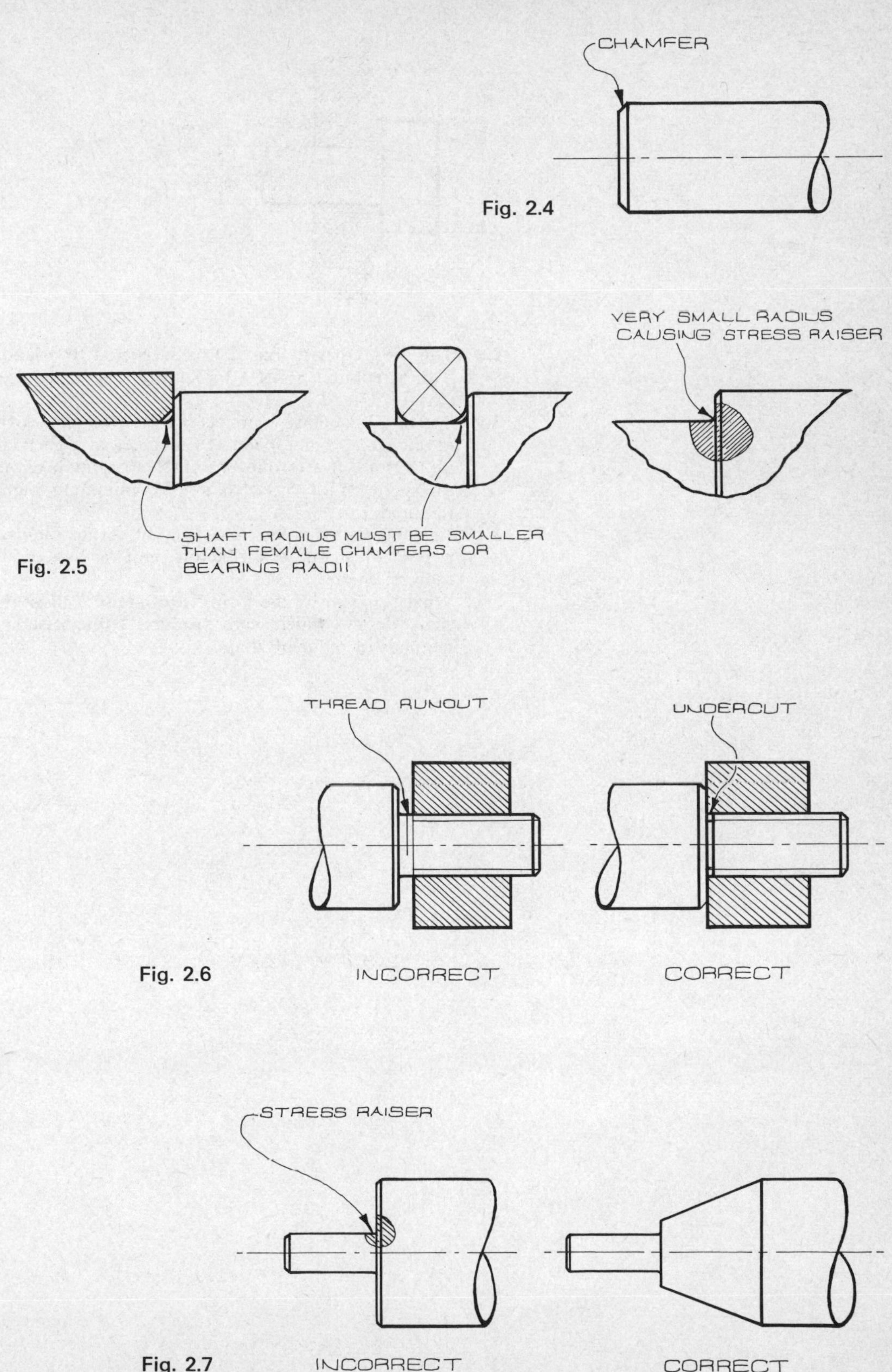

Fig. 2.4

Fig. 2.5

Fig. 2.6

Fig. 2.7

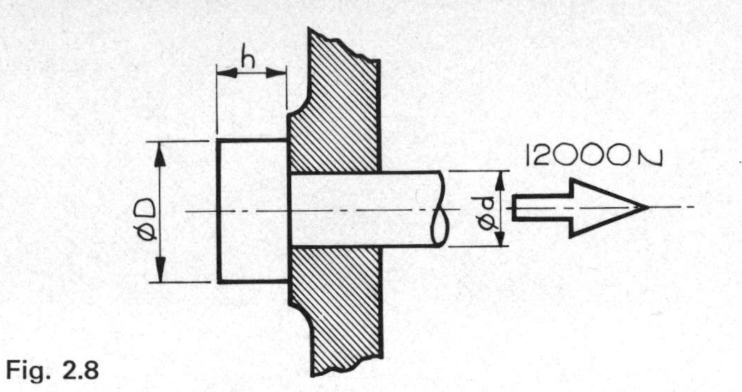

Fig. 2.8

2.8 This En1A bar of Fig. 2.8 is subjected to a load of 12 000 N with a required SF of 10. Determine suitable sizes for dimensions *d*, *D* and *h*.

2.9 A flanged coupling connects two transmission devices with pins of diameter 12 mm made of En3 steel, at a pitch circle diameter of 150 mm. If a torque of 2600 N m is applied with a required safety factor of 5, calculate the minimum number of studs required.

2.10 Explain why chamfers and internal corner radii will always be present on shafts. What factors limit the size of internal corner radii on shafts?

2.11 What is meant by the term "stress-raiser"? List the features of shaft design which can produce stress-raisers and suggest methods of reducing them.

3 Design Form Considerations

3.1 General Considerations

The design form has a major influence on how a component will behave in service. In many instances the designer must develop the form so that it has high resistance to bending and shear loads, and at the same time be of minimum weight. The weight is entirely dependent on the material selected.

Performance, reliability and structural adequacy should never be sacrificed in a misguided interest to use the lightest material available.

Because of the development of aluminium alloys which can withstand loads that at one time only steel would have withstood they are becoming increasingly popular in the fields of automobile, aeronautical and agricultural applications. Although initially more expensive than other materials, they can be more readily machined at higher cutting speeds, which in many instances results in a cheaper product overall. Further, they have a higher resistance to corrosion and if necessary can be protectively coated to improve corrosion resistance.

Cast steel has a greater strength than cast iron and is used where design loads are severe, and where the advantage of casting is required. Thinner sections can be used, and this combined with adequacy of form results in a high strength component. Due to the higher density of cast steel it is not recommended in applications where light weight design is a criteria. Malleable irons are often used in place of steel castings as they can offer high strength for a lower cost. Generally the high strength is obtained by sacrificing impact resistance. However, appropriate heat treatments can sometimes be used in order to maintain a balance between tensile strength and ductility. These irons fall into four main groups: Blackheart, Whiteheart, Spheroidal Graphite (SG Iron), and Pearlitic. The SG irons have been developed for applications where higher tensile strengths are required; the modulus of elasticity is in the region of 172 GN/m^2 as against 210 GN/m^2 for steel.

In Chapter 2 consideration was given to the various types of materials available to the designer. It is obvious from this that the task of selecting the right material for the application will not be easy. To make the task even more difficult, the shape, or form, of the component will influence the final selection.

The process of manufacturing a component from basic raw material to the final finished machined component consists of many stages. Every step must be closely controlled by quality control methods in order to ensure that the part will do the job correctly in service.

These processes can be broken down into two categories:

a) Primary process
b) Secondary process.

The basic or **primary process** is concerned with whether the basic form of the component should be cast, forged, rolled, extruded, etc. The **secondary process** is the further operation necessary to finish the part and may consist of bending, forming, grinding, etc.

Having established the desired form of the component, the designer must look at the processes actually available and ensure that the result will be at minimum cost, and that problems due to such things as porosity in the case of castings or stress cracks in the case of forgings do not arise subsequently.

3.2 Casting Metal

The **casting** process can be applied to iron, steel, aluminium, aluminium alloys, brass, etc. Each type of material has its own limitations in respect of section thickness, fluidity of casting, and shrinkage. The greater the shrinkage and the higher the temperature of the metal, the more difficult the problems encountered in achieving the sound casting.

Figure 3.1 shows a spoked wheel. The poor design shows the spokes to be straight, resulting in them being cracked after the part has been cast. The good design shows that, if they are curved, then this problem can be eliminated. The reason this occurs is that the hub spokes and rim tend to cool at different rates, hence setting up internal stresses. These stresses will still occur even with the curved spoke, but flexure will tend to prevent cracking. It is always a good idea to apply a stress-relieving treatment to the part after casting in order to reduce these internal stresses, because subsequent machining processes can introduce other stresses or allow the inherent stress to be "elastically" released.

Shrinkage cavities, or porosity, are voids within the casting brought about by allowing insufficient material to compensate for poor design of the various sections and intersections of webs, arms, etc. Uniform solidification of the casting is necessary. An ideal casting is produced when solidification begins at

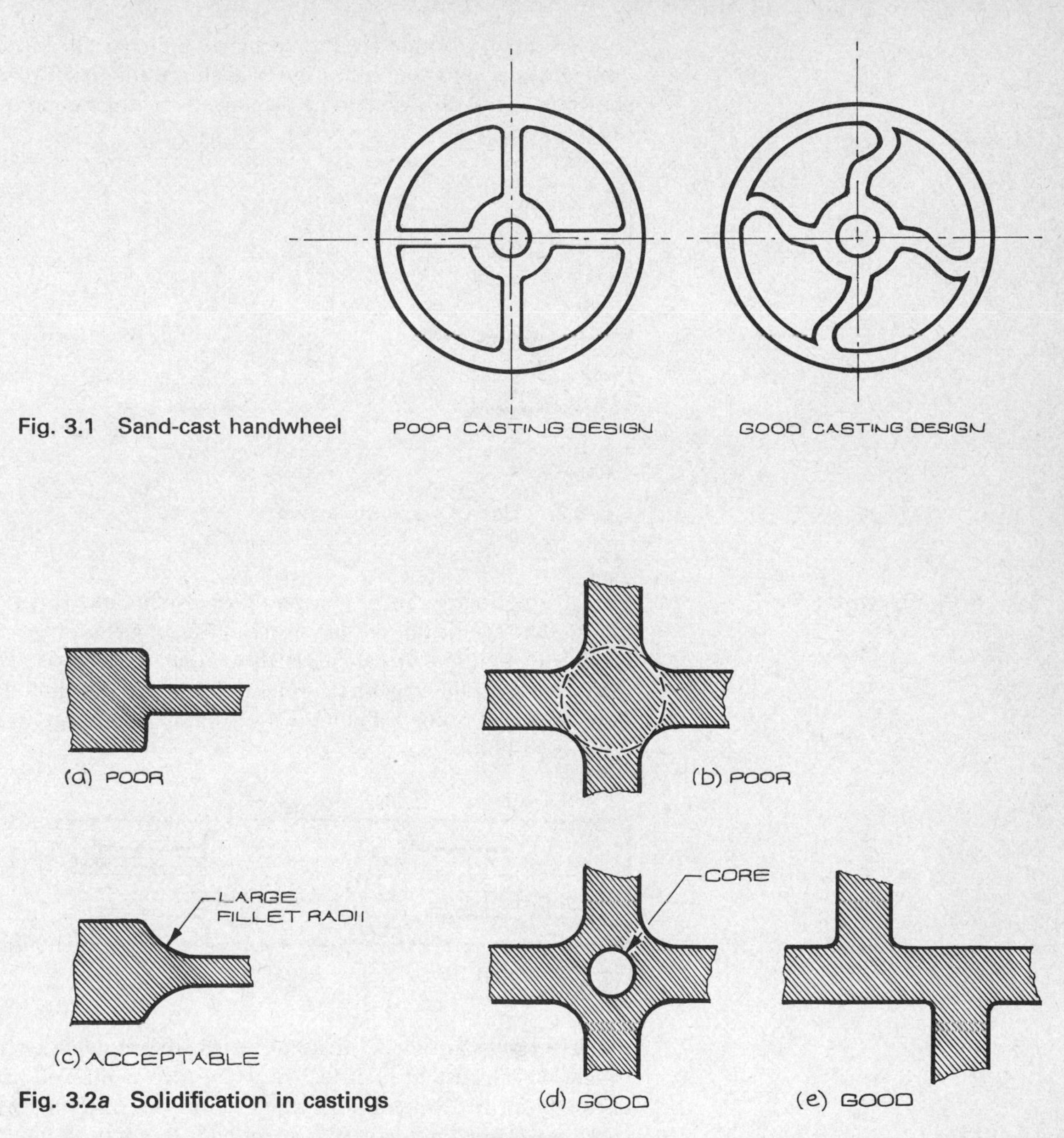

Fig. 3.1 Sand-cast handwheel

Fig. 3.2*a* Solidification in castings

the bottom of the mould and proceeds uniformly to the top. It is thus desirable that the casting thickness be kept uniform.

Figure 3.2*a* illustrates various methods by which uniform solidification may be achieved in design. Where these sections cannot be achieved, large fillet radii should be used. Large flat areas should be avoided as there is a tendency for warping. Webs or ribs should be added to support the flat area if they are unavoidable in the design.

Webs may also be used to achieve uniform thickness of casting (thus avoiding porosity) whilst retaining stiffness as shown in Fig. 3.2*b*. A further advantage here is the consequent reduction in weight and cost.

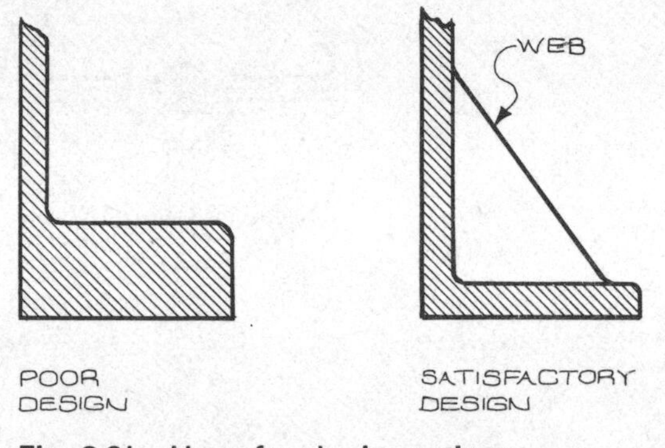

Fig. 3.2*b* Use of webs in casting

3.3 Forging Metal

Forged steel parts can produce a cheaper component depending on the form it takes. One of the most important aspects of forging in relation to casting is the ability to achieve refinement of the grain structure by the development of directional flow lines (grain flow). Figure 3.3 shows the typical grain flow for a gear blank.

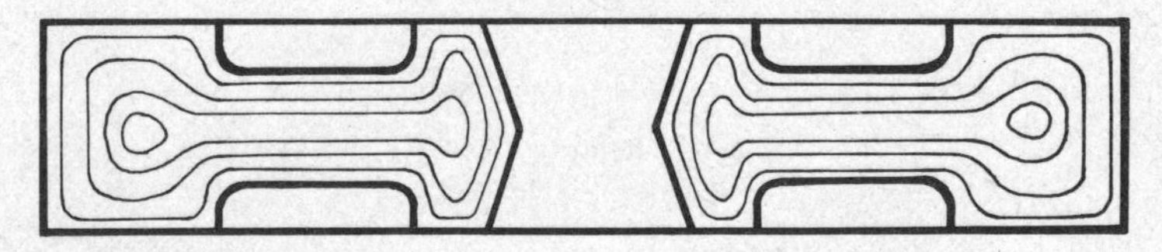

Fig. 3.3 Grain flow

Die forgings are made in steel dies which are made of hard, tough, wear-resistant and heat-resistant tool steel. The part is normally formed when the material is hot. Because the formed component must not stay in contact with the die too long, the process is quick and accurate and competes with alternative processes on high quality parts.

Forging processes include roll die and upsetting.

Roll die forging is the process of passing the material through rollers. The rollers are gradually brought closer to each other, which squeezes the material to shape. It is a very fast process, and is only economical on large quantity production. Closer tolerances can be held than with drop forging, being as low as ± 1 mm on small work, and $\pm 1\frac{1}{2}$ mm on large work.

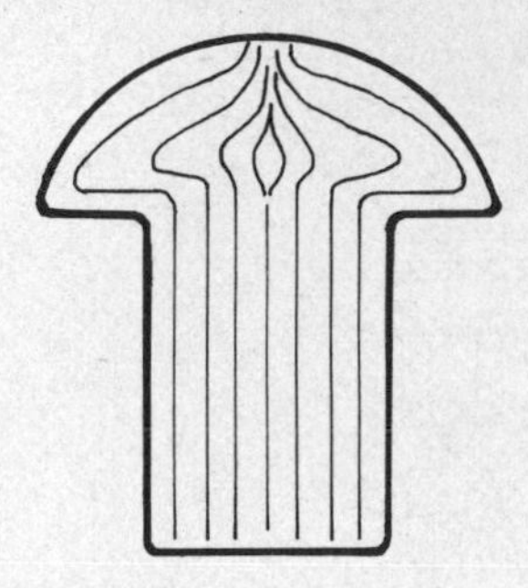

Fig. 3.4

Upsetting is similar to drop (or die) forging in that it is carried out with the use of dies. The forming of the head of a rivet from a small rod is an example (see Fig. 3.4). The head is increased in size by forcing material from the rest of the rod. This produces a regular grain flow which follows the contours of the form. This process, like hammer forging, results in increased strength and ductility.

Compared with castings, forgings have greater strength for the same weight. The initial cost of the forging die is usually greater than the cost of patterns, and care has to be taken to ensure that quantity and application will result in an economic component. However, the greater strength of a forging rather than the cost can sometimes dictate its choice over that of a casting.

3.4 Extrusion of Metal

Extrusion is the process of continuous-forming a material when in a heated condition. Great pressure is applied to a heated blank causing it to be passed through a "shaped" die. This process is normally confined to materials having a low melting point such as aluminium, copper, magnesium, lead, tin and zinc. Typical sections produced by this process are shown in Fig. 3.5.

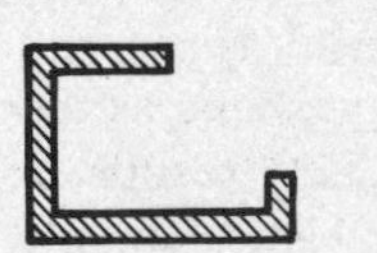

Fig. 3.5 Examples of extruded sections

Plastic articles can also be produced this way. Rods, tubes, strips can be extruded in the wide range of colours that are obtainable with thermoplastics. Costly finishing processes can be eliminated by this method. Many domestic products such as curtain rails, beading, picture framing can be produced by the process.

3.5 Rolling Metal

Rolling is a compression process and can be performed on hot or cold materials.

Cold rolling is a mill process which is carried out by passing hot rolled metals such as bars, sheets or strip through a set of rolls many times in order to reduce the thickness until the correct size is reached. Cold rolling increases the tensile strength and gives a bright finish to the part. Close tolerances can be held by this process.

Hot rolling refines the grain structure as opposed to distorting it in cold rolling. Hot rolling makes little change in the hardness or ductility. Large reductions in thickness can be achieved in a single pass and there will be no rupture or tears. The hot working process requires expensive tool steels for the dies and can result in high cost on small quantity production. Figure 3.6 shows how the grain is distorted when passing through rolls.

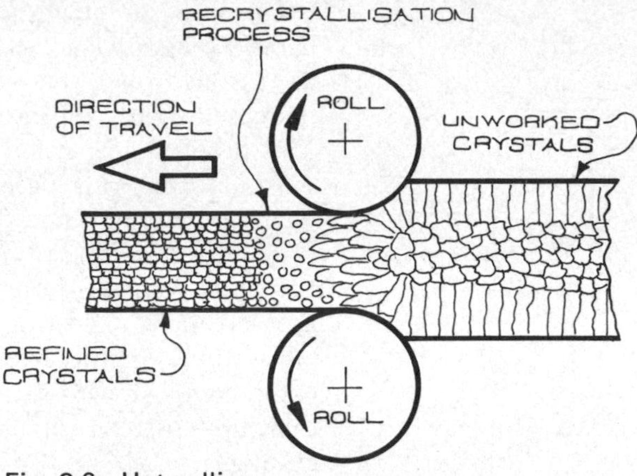

Fig. 3.6 Hot rolling

3.6 Plastics Moulding

The moulding of plastics material is not unlike the process of pressure die casting. The main difference is that the moulding material is softened by heat, instead of being completely molten.

3.7 Compression Moulding of Plastics

In **compression moulding**, the moulding material is introduced into a heated die and pressure is applied. The result is that the material is formed into one mass having the same shape as the die cavity. Further heating (thermosetting plastics) results in hardening of the moulding material. If thermoplastics are the material used in the moulding process then the hardening process takes place by cooling the die.

It is normal in any moulding process to use the die for only one type of material. This is because moulding materials tend to have different shrinkage rates, and different flow characteristics.

3.8 Injection Moulding of Plastics

Injection moulding is the process whereby the raw material, normally in the shape of granules or pellets, is placed into one end of a heated cylinder, then heated in a chamber, and finally pushed out of the other end of the cylinder, through a nozzle and into a closed mould. The component is then allowed to harden in the mould cavity. In order to speed the production of parts the heating chamber is designed to accept several changes of material based upon the mass of the finished component.

The pressure required in this process varies between 70 000 kPa and 175 000 kPa. Heating temperatures in the cylinder vary with the type of plastic material being used but are usually in the range of 93°C to 315°C. Thermoplastics require high barrel temperatures usually between 175°C and 315°C.

3.9 Transfer Moulding of Plastics

Transfer moulding is the process of forming articles in a closed mould, where the material is carried into the mould cavity under pressure from an auxiliary chamber. The material is placed in this hot auxiliary chamber and then forced in its plastic state through an orifice into the cavities by pressure. Heat and pressure must be held for a definite time to allow the chemical reaction (polymerizing or curing) to take place, depending on the sectional area of the piece.

3.10 Extrusion of Plastics

Extrusion is the process of continuous-forming a material by applying heat and mechanical working. This is the same process used in all extrusion processes, whether plastics or metals.

3.11 Casting of Plastics

Plastics can be **cast**, normally in the liquid state, to any contour that is practical in terms of the mould shape and the ability of the part to separate easily from the mould. The most common families of plastics used in this process are the epoxy, phenolic and polyester. The mould is usually made from lead, plastic or rubber latex.

3.12 Blow Moulding of Plastics

In **blow moulding**, a tube of melted plastic material is used and enclosed in a mould. This tube is then blown out to conform to the mould contour. Typical of this process are tubes for shaving cream, cosmetics and toothpaste.

This process can be integrated with the extrusion process to provide a continuous operation and speed production. Tube ends are closed by a subsequent heating process.

Air pressures between 250 kPa and 700 kPa are used. Usually the higher the pressure used, the better the surface finish. In order to control the wall thickness of the finished part, a low pressure of between 30 kPa and 150 kPa is used to blow the molten material to its initial form. The material is held against the walls of the mould by pressures of between 415 kPa and 700 kPa so that it can cure or minimise wrinkling and distortion.

3.13 Welding of Metals

Welding is a metal-joining process in which two components of like composition are melted and fused together, sometimes with the aid of a filler metal. The result is a permanent joint which, given good design, will be as strong and rigid as any equivalent bolted or riveted assembly. Welding is most commonly undertaken in steel fabrications but is also widely used with cast irons and non-ferrous metals such as aluminium.

Welding may be broadly split into two groups;

a) Fusion welding
b) Pressure welding.

3.14 Fusion Welding

In **fusion welding**, the metals fuse together entirely, by the application of heat and filler metal at the joint. Heat is supplied by oxy-acetylene gas ignition or by electric arc, sometimes using specialized gas-shielded techniques. In the case of electric arc welding, the flux-coated filler metal forms the electrode. There are many different types of fusion-welded joints, the most common of which are illustrated on page 138. Also shown on page 138 are the British Standard Symbols (BS 499 Part 11) for these joints as they should appear on an engineering drawing.

Fillet welds These provide a cheap weld for right-angled sections. Fillet welds are the most common type of joint and require no edge preparation of parent metal.

Bevel butt welds (Single or Double) These are also used for right-angled sections and will withstand greater mechanical loading than fillet welds. The double bevel butt provides the strongest right-angled joint. Since the edge of one parent surface has to be shaped, bevel butt welds are more expensive than fillets.

Square butt weld These provide a cheap weld for thin plate surfaces (up to 4 mm thick) which are in line. No edge preparation is required.

V-butt weld (Single or Double) These are used for in-line sections above 4 mm thickness. Parent surface edges are shaped to give a strong joint. The double-V butt weld provides the strongest in-line joint.

3.15 Pressure Welding

Pressure welding methods produce welded joints by applying both heat and pressure to the parent surfaces. The most widespread pressure welding technique is called resistance welding, which forms the basis of spot welding and seam welding processes.

Spot welding This process is widely used for lapping together thin metal sheet (Fig. 3.7) and lends itself particularly to mass production techniques for applications such as car bodywork. The process is quick, clean, cheap in large quantities, and requires no filler metal.

High current is passed through two electrodes which are applied to the metal at pressure. Heat and fusion at the interfaces are produced by the resistance across the air gap between the metals.

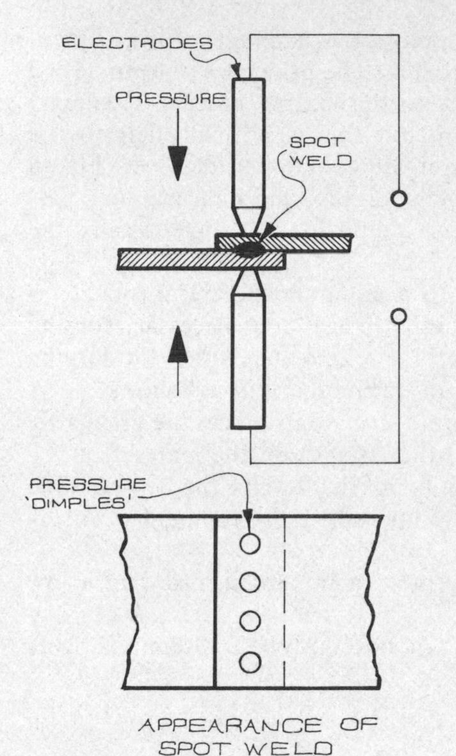

Fig. 3.7 Spot welding.

Seam welding This is a similar process to spot welding, but the electrodes are in the form of revolving wheels, in order to aid mass production and create an air-tight joint.

Steel welded structures These may be made up entirely of plate or may be a mixture of rolled-on drawn sections such as rectangular or circular bars, tubes, angles, tee sections and RSJs. As with a cast component, the welded fabrication may require subsequent machining.

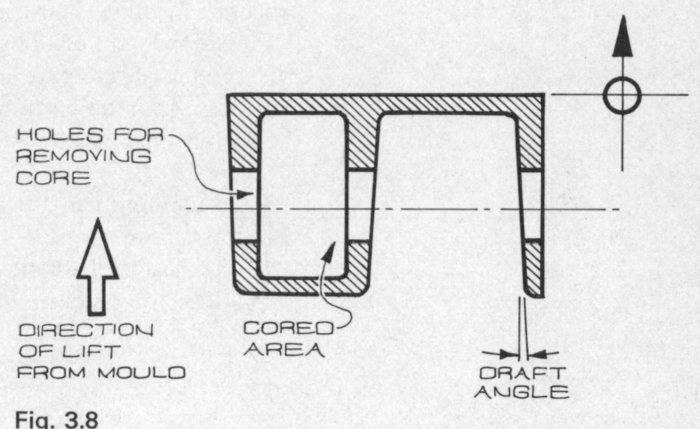

Fig. 3.8

3.16 The Design Form

It is necessary that, whichever process is adopted to produce a component, the shape or form of the part follows certain basic rules. These rules will depend on size, material, load-carrying criteria, and cost limitations.

Casting is in itself a complete process whereby intricate shapes can be produced. However, in many instances, in order to produce the intricate shapes, cores have to be made which can only be removed from the completed casting by breaking them. Figure 3.8 shows a typical cored area of a casting which requires such breaking.

Wherever possible, draft angles must form part of the design and be designed so that the casting can be lifted from the mould without the risk of tearing away the moulding sand. These draft angles must be positioned in the direction as shown in Fig. 3.8.

Split lines, as indicated in Fig. 3.9, must be positioned by the designer to ensure that the mould design is kept as simple as possible. Figure 3.9 shows a hub for a vehicle axle bearing carrier and the recommended split line is shown.

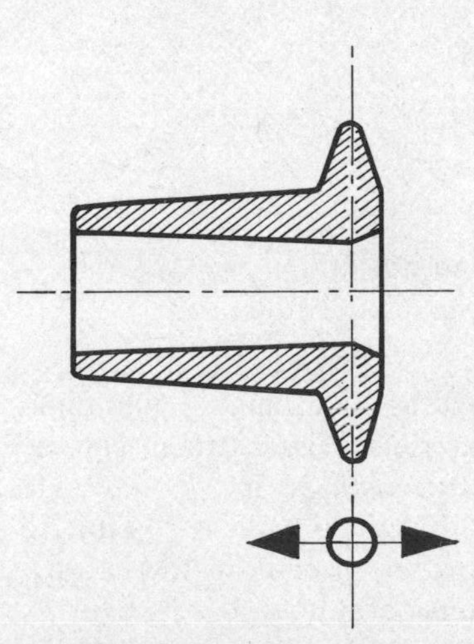

Fig. 3.9 Split line

3.17 Forging Design

Forging is a process whereby the amount of subsequent machining is kept to a minimum. The principles of form design is in many respects similar to the casting process. Adequate draft must be designed in, but this is normally left to the forging die designer and not the component designer. Figure 3.10 shows a drive shaft gear forging, and indicates how part of the design can incorporate a built-in draft angle to assist the die designer.

Machining can be kept to a minimum if careful thought is given to the part at the design stage. It is necessary for the designer to be fully aware of the advantages that the forging process can offer. These can be summarised as follows:

1. Compaction of the grain structure reduces the grain size, giving an increase in the strength of the material.
2. The grain flow can be controlled to give the best possible resistance to stresses imposed by the component in service.
3. Machining can be kept to a minimum, ensuring a low cost component.
4. Accuracy of shape can be controlled within relatively close tolerances.

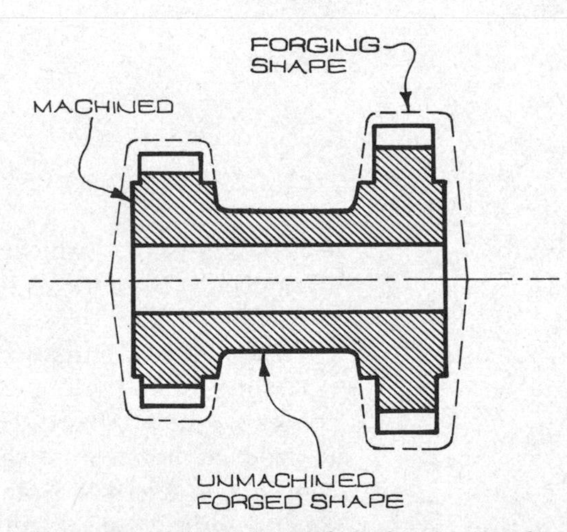

Fig. 3.10 Adequate draft in forging design

3.18 Die Casting Design

Die casting is normally confined to aluminium, aluminium alloys, zinc and magnesium materials. The castings can be produced more accurately than sand castings, and good die life can be achieved on large quantity production. It is possible to cast small diameter holes and threads directly in the casting, thus reducing machining costs considerably.

When die cast parts are made of ductile material, they can

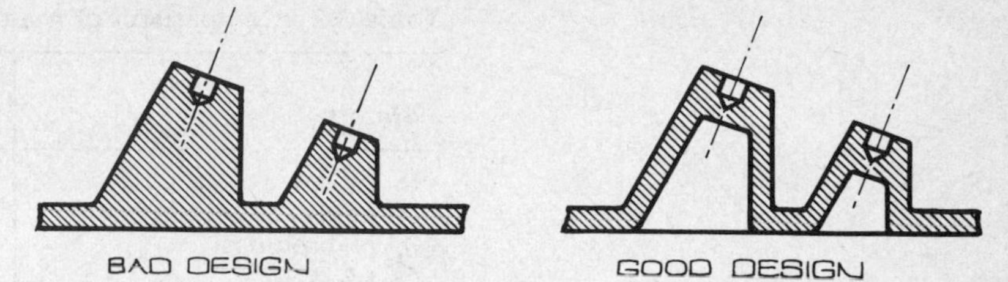

Fig. 3.11 Casting around cored holes

be bent, formed or joined to another piece after casting. Rivets, bolts, etc. can be cast in to the part, which eliminates subsequent staking or tapping operations.

As with any casting process, good design practice dictates that the wall thickness is kept uniform. Walls must be thick enough to permit proper filling but thin enough for rapid chilling in order to obtain maximum mechanical properties. Figure 3.11 shows a typical design of casting around cored holes, indicating good and bad design practice.

It should be remembered that, if holes are to be cored, standards exist which limit the ratio of hole diameter to depth of hole. Generally this ratio is approximately 4 to 1. That is, for a 5 mm diameter hole, the depth should not exceed 20 mm. However, it is advisable to discuss casting requirements with

Table 3.1 Typical values of tolerances in die casting

	Zinc	*Aluminium*	*Magnesium*
Basic tolerance (up to 25 mm)	±0.075	±0.100	±0.100
Additional tolerance (25 to 300 mm)	±0.025	±0.035	±0.035
Additional tolerance (over 300 mm)	±0.025	±0.025	±0.025

the foundry design engineer at an early stage of the design.

The tolerance for dimensions of die casting varies depending on the supplier. However, Table 3.1 gives some typical values. These tolerances depend on the accuracy to which die cavity and cores are machined, the thermal expansion of the die during casting, the tolerance on the position of separate parts of the die, and the surface finish of the die (1.2 to 32 microns are fairly common for this feature).

3.19 Economic Considerations

It is the designer's job to compare the costs of the various materials suitable to achieve the required physical and mechanical properties. Table 3.2 shows how the costs of various materials differ. The table also shows typical densities for each material. This gives a good indication of the weight-saving that can be achieved.

Table 3.2 Comparison of material costs

	Material	*Density* g/cm³	% cost/kg	% cost/cm³
PLASTICS	Vinyl	1.30	147	25
	Polystyrene	0.92	156	19
	Fibreglass	1.61	226	47
	A.B.S.	0.83	339	36
	Nylon	1.14	739	109
IRONS	Grey iron casting	7.10	139	128
	Ductile iron casting	7.10	278	256
	Malleable iron casting	7.30	330	313
	Steel casting	7.91	348	356
DIE CASTINGS	Zinc die casting	6.66	365	316
	Aluminium die casting	2.66	478	166
	Brass sand casting	8.88	913	1046
	Manganese bronze casting	8.21	1391	1481
STEEL	Sheet steel (En1A)	7.85	59	59
	Bar steel (En1A)	7.85	100	100
	Steel forging	7.85	304	309

It is a good plan for the designer to make a tabular comparison in the first instance so that the most economic choice can be made.

The following is an example of how this could be done.

Example

Consider the possibility of producing a machine part to the design shown in Fig. 3.12. Establish the total cost of producing the part as a

a) Grey iron casting
b) Aluminium casting
c) Steel drop forging

What would be the cost of producing 1000 components like this?

Solution The first step is to calculate the volume of material used. For the purpose of this example we will assume it to be 100 cm³.

Next make a table as shown in Table 3.3.

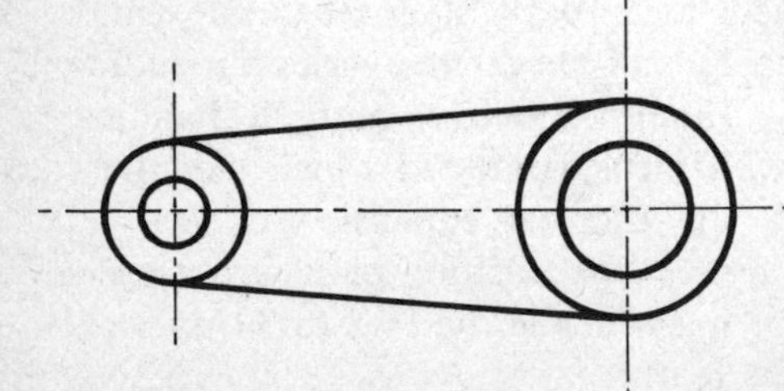

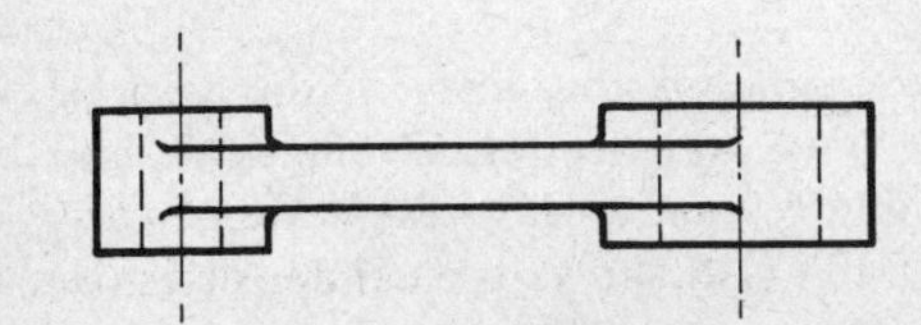

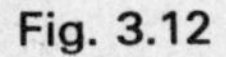

Fig. 3.12

EXPLANATION OF TABLE

Line A This is the cost of the basic material. If only one material is known, for example the cost per kilogramme of

Table 3.3

Cost/unit	Grey iron casting	Aluminium die casting	Steel forging (drop)
A Cost of material per kg	20p	70p*	44p
B Total material cost (100 cm^3 volume)	14.2p	18.62p	34.54p
C Cost of pattern or die	£50	£350	£375
D Labour cost for blank	20p	70p	65p
E Labour cost for machining	£1.77	70p	£1.20
F Manufacturing cost for one piece part	£52.112	£351.59	£377.20
G Neglecting pattern or die cost	£2.112	£1.59	£2.20
H Amortised over 1000 piece parts including pattern or die cost	£2162	£1936.2	£2570
J Cost per piece part	£2.16	£1.94	£2.57

grey cast iron, then the other material costs can be estimated from Table 3.2.

Line B This is the total cost of the material for the part being considered. It can be calculated by using the density values in Table 3.2 thus:

For grey cast iron, the cost is

$$7.10 \times \frac{100}{1000} \times 20 = 14.2 \text{ pence}$$

For aluminium die casting, the cost is

$$2.66 \times \frac{100}{1000} \times 70 = 18.62 \text{ pence}$$

For the steel drop forging, the cost is

$$7.85 \times \frac{100}{1000} \times 44 = 34.54 \text{ pence}$$

Line C This is the estimated cost of producing the pattern or die. The value will vary for each manufacturer and it is the buyer's function to obtain the best possible quotation.

If it is found to be difficult to obtain a realistic cost at the design stage, then sometimes it is possible to take a calculated guess based on the designer's own experience, or based on one of the other products in the factory.

Line D The labour cost of producing the blank depends on the size of the company which is going to produce the part. A small company will not have such high overheads to account for. However, a reasonable estimate should be possible if the foundry is contacted at an early stage of the design.

Line E The labour cost of producing the machined part in the factory is a relatively easy one to obtain. The company planner and cost estimator should be contacted at this stage and they should be in a position to supply this figure, given the basic design form.

Line F The manufacturing cost for one part is the value obtained by adding together **B**, **C**, **D** and **E**. The high cost here arises from the pattern or die cost. If only one part is required it can be seen in this example that the grey iron casting would be the cheapest method of production.

Line G The manufacturing cost neglecting the pattern or die cost is shown here, but in this instance has little significance except that at this stage the designer may want to consider other possible methods of production, i.e. welded fabrication. However, the labour cost of producing the blank may make this an expensive choice as it can sometimes be 5 or 6 times more expensive than for a grey iron casting.

Line H This is the total estimated cost to produce 1000 parts, and is given by

$$[1000(\mathbf{B}+\mathbf{D}+\mathbf{E})+\mathbf{C}]$$

Therefore, for grey iron casting, the amortised cost is

$$[1000(14.3+20+177)+5000]$$
$$=216\,300 \text{ pence}=£2163$$

For aluminium die casting. the amortised cost is

$$[1000(18.62+70+70)+35\,000]$$
$$=193\,620 \text{ pence}=£1936.2$$

For steel drop forgings, the amortised cost is

$$[(34.54+65+120)\times 1000+37\,500]$$
$$=257\,000 \text{ pence}=£2570$$

Line J The cost per part is calculated by dividing line **H** by 1000.

In conclusion it can be seen that, for this design cost exercise, it would be most economical to produce the part as an aluminium die casting. It is 10% cheaper than a grey iron casting and 25% cheaper than a steel drop forging.

It must be remembered that this will not always be the case and only by doing a cost estimate for each individual component part can the designer establish the most economic method of manufacture.

Always check the material cost at the time the design is nearing completion, as material costs can vary considerably over the period between design conception and manufacture.

3.20 Summary

The design form will depend on

1 Application conditions
2 Design loads
3 Weight criteria
4 Manufacturing criteria
5 Cost.

The above sections have shown some of the typical considerations that must be given to the design of a component. This is applicable whatever material is chosen, but it should be remembered that each material has its own limitations.

Always choose a material that will give the best cost-to-weight ratio when compared with the necessary strength criteria due to the applied design loads.

Where shock loads are expected, avoid using a material or form that gives rise to stress-raisers. The notch sensitivity (i.e. low impact resistance) of a material is a major consideration when designing the component part.

Design for uniformity of wall thickness, avoid sharp corners, and above all choose the right form from the strength aspect. When designing a casting, ensure that the simplest method of core extraction has been used. Avoid enclosed cavities, and always apply adequate draft angle to the casting shape to ease extraction from the mould.

Problems

3.1 Figure 3.13 shows the design for a bearing pedestal made from cast iron. Establish the cost of producing the part as

a) Grey iron casting
b) Aluminium die casting
c) Steel casting.

Re-design the shape of the part to suit each type of casting process. Using Table 3.2 establish the material cost of aluminium die casting and steel casting if the cost of grey cast iron is 20 pence per kilogram.

Estimate realistic costs for the pattern or die and for the labour cost for blank and its machining.

3.2 Give typical examples of every-day components that can be manufactured by

a) Metal extrusion process
b) Plastics moulding process.

3.3 Design an electrical wall switch which can be manufactured by plastics moulding techniques. Show how you would split the mould to simplify the removal of the part.

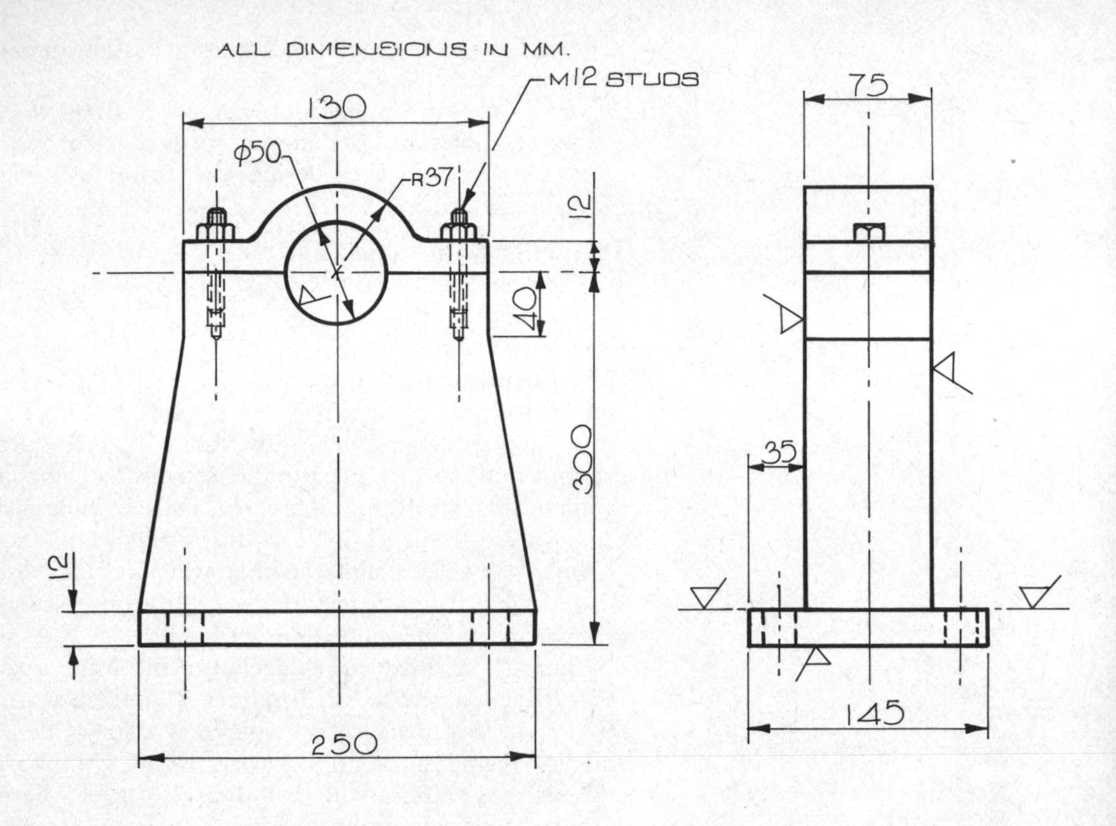

Fig. 3.13

3.4 For each of the fabricated components shown on page 45, identify the type of weld and draw neat sketches looking in the direction of the arrows shown. On each sketch show the type of weld using the appropriate British Standard symbols from Chart 1 (page 138).

3.5 Define the term "resistance welding." Name and describe one common type of resistance welding process. List three common articles in which this process is employed.

3.6 With the aid of sketches, give some common examples of bad casting design and describe the effect on mechanical properties. In each case explain how the design may be improved.

3.7 List the advantages of forging metal components. Describe two common forging processes.

3.8 Compare the processes of hot-rolling and cold-rolling of metals, listing the advantages and disadvantages of each. Name a typical component produced by each process.

3.9 Describe *a*) the injection moulding process, *b*) the blow moulding process, as applied to plastics materials. Name three common domestic articles which could be manufactured by each process.

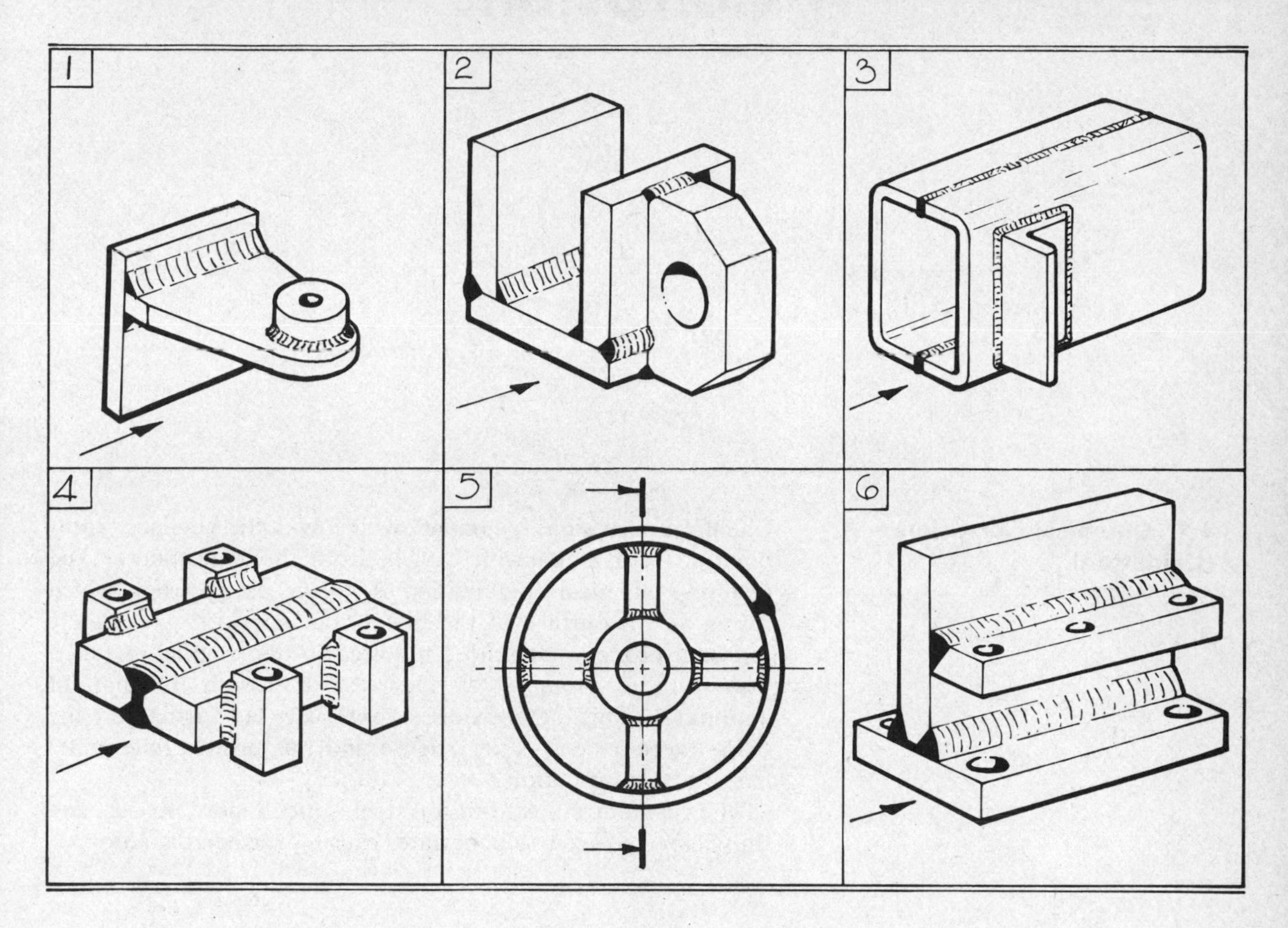

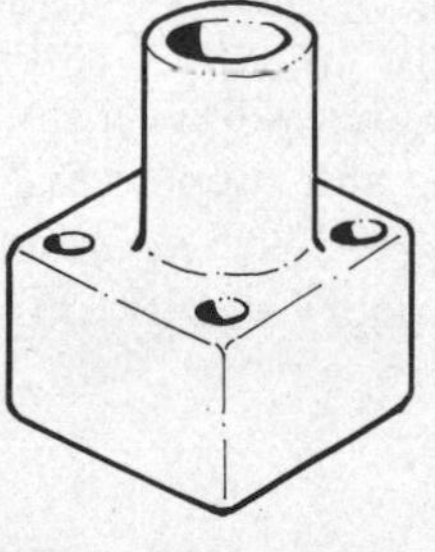

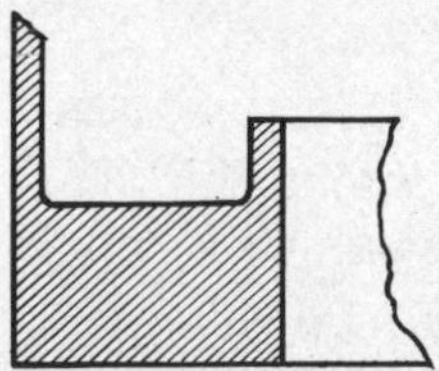

Fig. 3.14

3.10 A typical type of car door handle is made from zinc alloy, die-cast, chromium-plated, and machined in parts. In similar fashion, describe typical examples of the following items:

a) carburettor body
b) crankshaft
c) electric plug pin
d) hexagon bolt
e) lathe base-plate
f) connecting rod
g) structural angle section
h) section of metal window frame
i) domestic metal water pipe
j) motor cycle crankcase

3.11 Figure 3.14 shows two examples of bad castings. Improve these casting designs from the points of view of shrinkage porosity and cost.

4 Corrosion

4.1 Chemical Corrosion (Oxidation)

Chemical corrosion, or **oxidation**, involves the chemical combination of a metal with oxygen from the atmosphere. The resulting chemical combination is known as an oxide which forms on the surface of the metal and may have completely different properties to those required in the original metal.

A typical example of chemical corrosion is that of aluminium. This metal oxidises very easily but fortunately the oxide layer formed is very dense and this protects the metal from further oxidation.

The rust film which forms on steel is much more porous, and thus allows oxygen to penetrate, causing further oxidation.

4.2 Electro-chemical Corrosion

Electro-chemical corrosion is best illustrated by the principle of the simple battery cell. A battery is set up when two different metals are close to each other and covered with an electrolyte liquid. When they are connected by a conductor, an electric current flows which results in one plate (*anode*) dissolving in the electrolyte and gas bubbles being deposited on the other plate (*cathode*). For example, using zinc and copper (Fig. 4.1), the zinc will always dissolve because zinc has *negative* potential towards copper (or is anodic to copper) and its ions are attracted towards the copper cathode.

Gold	+1.68
Platinum	+1.40
Silver	+0.80
Mercury	+0.79
Copper	+0.52
(Hydrogen)	0.00
Lead	−0.13
Tin	−0.14
Nickel	−0.21
Cadmium	−0.40
Iron	−0.44
Chromium	−0.56
Zinc	−0.76
Aluminium	−1.67
Magnesium	−2.38
Sodium	−2.71

ELECTRO-CHEMICAL SERIES

Metals may be arranged to indicate their potential towards each other. One which is more negative than another is anodic and it would thus corrode in close contact.

For example, magnesium is very anodic to silver and would corrode rapidly with the two close together in an electrolyte.

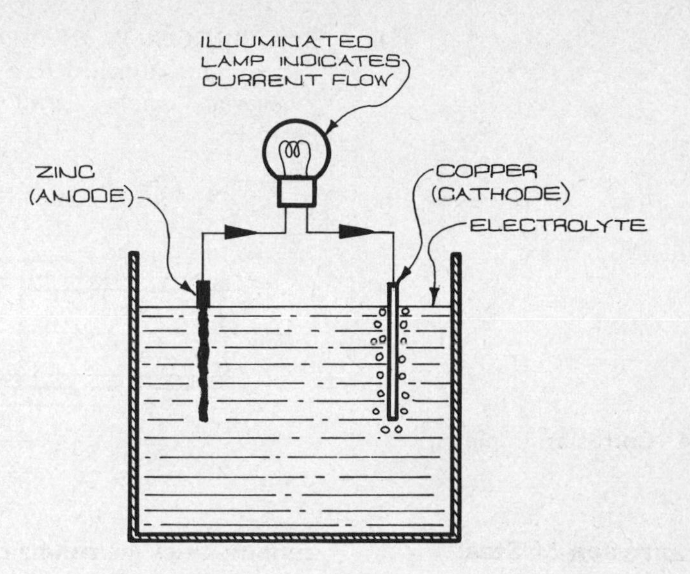

Fig. 4.1 A simple battery

4.3 Examples of Electrolyte Corrosion

1 SCRATCHED TIN PLATE (Fig. 4.2) Tin is coated on steel to protect it from oxidation but, once the film is scratched, then with steel anodic to tin and with moisture in the air as an electrolyte, the steel corrodes quicker than with no coating at all.

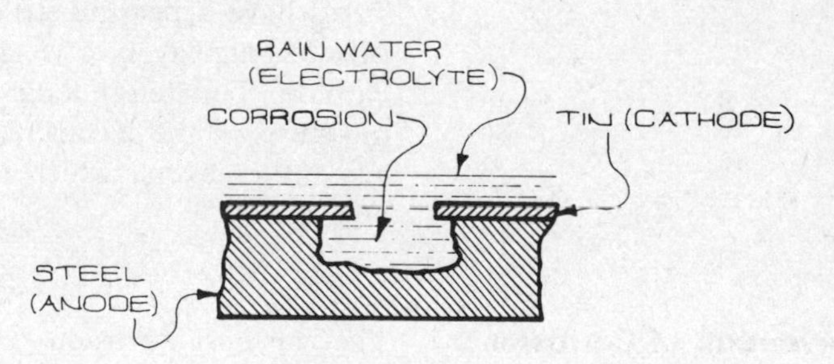

Fig. 4.2 Scratched tin plate

2 SACRIFICIAL PROTECTION OR GALVANISING (Fig. 4.3) **Galvanising** involves giving mild steel a zinc plate coating. If the zinc is scratched, then with zinc anodic to steel, the zinc corrodes instead of the steel.

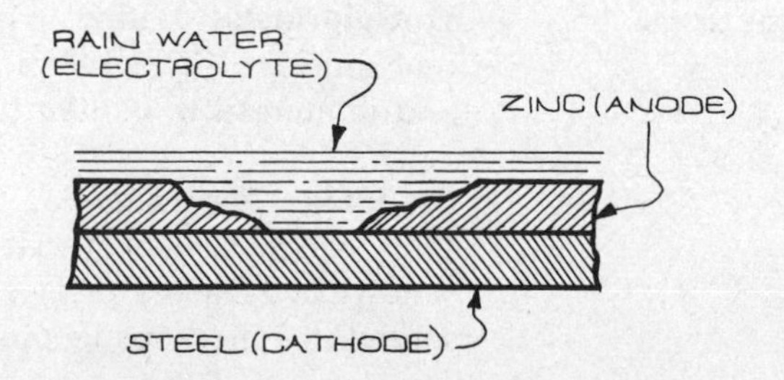

Fig. 4.3 Galvanising

3 CORROSION IN PIPING (Fig. 4.4) The situation of a copper pipe attached to a mild steel pipe should be avoided wherever possible as steel is anodic to copper and thus quickly corrodes.

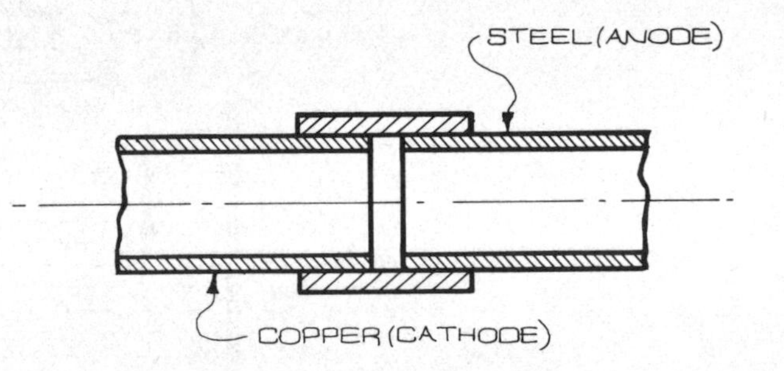

Fig. 4.4 Corrosion in piping

4.4 Corrosion of Steel (Rusting)

In most cases the **rusting** of steel is a mixture of chemical and electro-chemical corrosive action. The iron and the steel chemically combine with oxygen in the air to form a porous oxide layer on the surface of the metal. The corrosion process is aided by the fact that iron is anodic to the oxide. Electrolytic corrosion thus takes place with moisture in the porous oxide acting as the electrolyte.

Electrolysis can also occur inside the metallic crystals. Many steels have a pearlite structure, in which each crystal consists of alternate layers of ferrite (pure iron) and cementite (iron carbide). Ferrite is anodic to cementite and thus corrodes, with moisture as the electrolyte (Fig. 4.5). This leaves the brittle cementite layers, which ultimately fracture as corrosion increases.

4.5 Prevention of Corrosion

The various corrosion protection methods may be broadly listed under the following headings:

1. Sacrificial protection (galvanising).
2. Use of two metals close to each other on the electro-chemical series.
3. Protective coatings.

4.6 Protective Coatings

Hot-dip metal coating For example, tin-plating, zinc plating (galvanising). Steel is "pickled" (chemically cleaned in acid) and immersed in molten tin or zinc.

Cladding For example, Alclad (aluminium coated duralumin), Niclad (nickel coated steel). The parent metal is sandwiched between pieces of the coating metal and the sandwich rolled out to the required thickness.

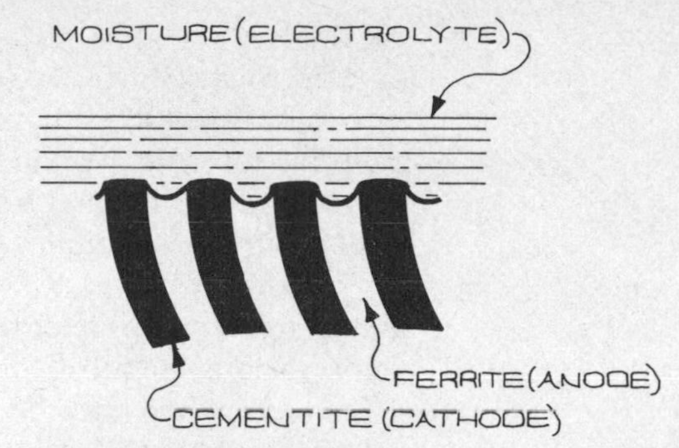

Fig. 4.5 Corrosion of pearlite

Sherardising Like galvanizing, this process involves the zinc-coating of steel. Zinc is made to chemically combine with steel by heating the work with "zinc dust" at a temperature below the melting point of zinc. Much thinner coating may be obtained than with galvanising.

Electroplating In all electroplating processes, a protective coating is applied by electrolysis action. Typical coatings include copper, nickel, chromium, cadmium, gold, silver, tin, zinc. A high accuracy of coating thickness can be readily obtained.

Oxide coatings The oxidation of some metals produces a very dense oxide layer, which itself prevents further oxidation. In some processes it is possible to exaggerate this natural process to produce an efficient protective coating. A good example of one of these processes is **anodising**, in which natural aluminium oxide is thickened. The aluminium is used as an anode in an electrolyte process. When an electric current is passed, oxygen forms on the aluminium and immediately combines to form the oxide layer.

4.7 Surface Preparation Methods

Cleaning methods are numerous in the metal industries, and are used as a means of removing dirt, oil, oxide scale and other harmful elements that would ultimately impair the operation of equipment, and the life of subsequent protective finishes.

These methods are not cheap to implement, and it should always be remembered that the right method should be adopted to suit the material, and be effective on the element that is to be removed. Considerable expertise is therefore necessary to remove this unwanted surface contamination.

Chemical processes such as pickling, and solvent, alkaline, electrolytic, and emulsion cleaning are used. Other more sophisticated treatments are now available such as ultrasonics, which can be very effective on mass-produced parts.

Pickling Castings, rods, sheets, bars, etc., particularly those of a ferrous nature, are often coated with a film of oxide, especially when they have been hot rolled. To prepare the material for cold working, this layer of oxide must be removed, and the method frequently used is to pickle the iron or steel in an acid solution.

This chemical process removes the scale and also has the ability to show up defects such as cracks and seams. Hydrochloric acid is the most commonly used—and is used cold.

Stainless steels may be pickled in nitric acid since hydrochloric acid can be harmful to the steel if the cold working is done too soon after pickling.

Sand and shot blasting To clean castings from scale, sand blasting can be used in which sand is projected onto the surface by an air jet or other blast.

Shot blasting is similar, except that the particles are chilled iron shot. This process is considered to have a beneficial effect on castings as the intensity of the shot hammers the surface and relieves internal stresses in the material.

Electrolytic cleaning An alkaline cleaning solution is used in electrolytic cleaning, and an electric current is passed through the solution which causes the emission of oxygen at the positive pole and hydrogen at the negative pole. This action breaks up the oil film holding the dirt to the surface. The process is mainly used where the component is to be plated.

Grinding Large iron castings may be de-scaled by a surface hand grinding wheel tool. This is a slow and arduous process and an extremely dirty one for the operator. Subsequent shot blasting is recommended to clean the surface thoroughly if possible.

4.8 Surface Coatings and Treatments

Surface protective coatings for metals were discussed earlier. However, other types of finish are available to the engineering designer in order to provide a decorative as well as a protective finish to the metal surface. These consist in the main of organic and inorganic coatings.

4.9 Organic Coatings

These include oil paint, enamels (baked on), varnishes, lacquers, etc., and may be applied by brushing, spraying, dipping or tumble finishing.

Oil paint This process completely hides the surface, and a variety of colours can be obtained to enhance the appearance of the finished product. Care has to be taken to prevent the possibility of the parts becoming adhered to each other during the drying process.

Enamels These are normally applied using a "baking-on" process to ensure that they effectively adhere to the metal surface. This provides a finish that is harder and more abrasion-resistant than many of the other types of finish. The automotive, electrical and domestic appliance industries are the main users of this treatment.

Varnishes These can be clear or opaque, and may contain a dye. They are normally fast drying and produce a glossy hard film which can have a toughness and durability suited for electrical equipment. The alkyed types of varnish have good adherence to smooth surfaces. Their exterior durability is very satisfactory, making them ideally suited for vehicle finishes.

Lacquers These are quick-drying finishes. However, their main drawback is the lower coverage possible compared to an equal unit of paint, varnish and enamel, and at least two coats are required compared with one coat of varnish or enamel. The durability of the finish is poor in comparision with the other finishes and they are therefore not so popular in the engineering industries.

4.10 Inorganic Coatings

These are made up of refractory compounds. They are harder, more rigid, and have greater resistance to elevated temperatures than organic coatings. They give eye-appealing finishes and offer good resistance to corrosion.

Inorganic coatings include the porcelain, enamels and ceramic coatings. They may be applied to ferrous and non-ferrous surfaces. The biggest drawback is their high cost and they should only be used in special cases.

Porcelain Porcelain enamelled surfaces give a good strength and stable finish and beautify the surface to make the product a more saleable item. Porcelain enamels will resist temperatures up to 550°C. Coating thicknesses are in the region of 0.075 to 0.100 mm.

Ceramic coatings These are vitreous and metallic oxide coatings that are more refractory than the porcelain enamels. In addition to protecting the metallic surface from oxidation and corrosion, it also increases the strength and rigidity. This last characteristic is important when the part is subjected to high temperatures.

Problems

4.1 Describe a common example of a chemical corrosion process.
4.2 Explain the meaning of "electrolytic corrosion".
4.3 Tin-plated steel is often used in the design of food containers because tin is non-poisonous. Explain why, in most other cases, zinc-coated (galvanized) steel is a more desirable method of corrosion prevention.
4.4 A common design fault in domestic water systems is the connecting of copper piping to a galvanized tank. Explain why this is most undesirable and suggest an alternative design scheme.
4.5 Describe the common corrosive action of steel (rusting).
4.6 What do you understand by the term "pearlitic corrosion".
4.7 Name and describe three common methods of preventing corrosion by the use of protective coatings.
4.8 "Pickling" and "electrolytic cleaning" are popular methods of surface preparation for metals. Describe them.
4.9 Compare the general advantages and disadvantages of organic coatings with those of inorganic coatings on metal surfaces. Give two examples of each.
4.10 Describe and compare the following types of organic coatings: oil paint, enamels, varnishes, lacquers.

5 Gears

A gear may be defined as a component which transmits rotary motion via the action of meshing teeth (Fig. 5.1). The meshing gears are termed **pinion** (for the driving gear) and **gear wheel** (for the driven gear).

A gear system may be incorporated into a design for the following reasons:

1 To change the speed of a machine.
2 To change the direction of rotation of a machine.
3 To increase the torque capacity of a machine.

5.1 Tooth Forms

A meshing tooth action will inevitably be a combination of rolling and sliding. The sliding part of the action produces frictional losses and therefore inefficiency in the machine. The principal objective of the gear design is to minimize the sliding effect. Centuries of evolution have transformed gears from crude square sections on wooden wheels into the most efficient modern tooth form which is based on a curve known as the involute (Fig. 5.2).

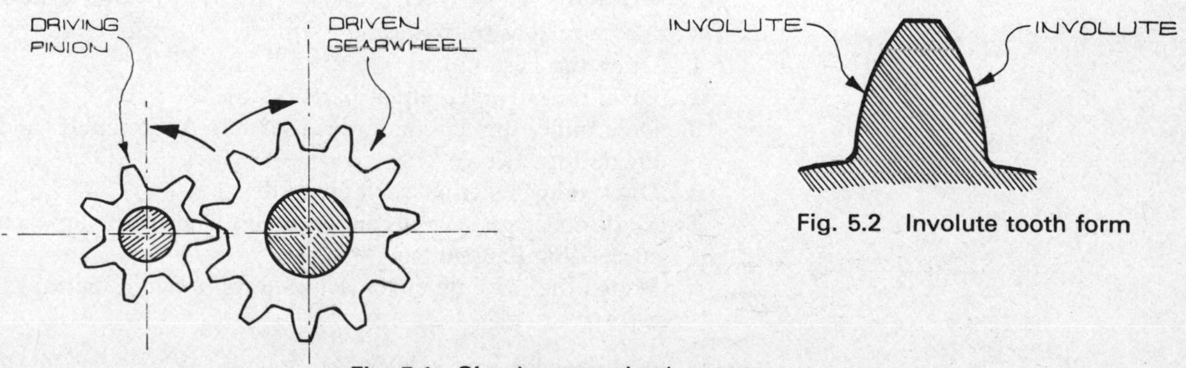

Fig. 5.2 Involute tooth form

Fig. 5.1 Simple gear wheel system

5.2 The Involute Curve

An **involute** is generated by a straight line rolling around a circle without slipping. The points on the straight line produce the involute curve as the line unwinds. The circle about which the line unwinds is called the *base circle.* At any point of unwinding, the straight line is of course a tangent to the base circle.

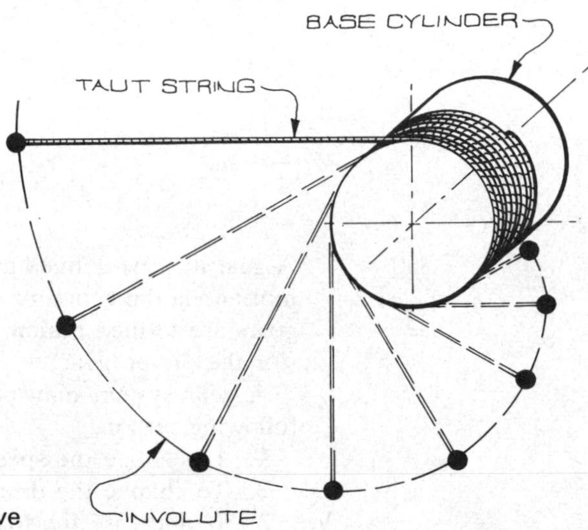

Fig. 5.3 Construction of involute curve

5.3 Construction of Involutes

A practical method of construction is to wind a piece of string around a base cylinder and then unwind keeping the string taut (Fig. 5.3). The string is then equivalent to a rolling tangent on a base circle, and thus traces an involute curve. Each successive tangent length will be equal to the unwound circumferential arc length.

The professional draughtsman will of course require geometric methods of construction.

GEOMETRIC CONSTRUCTION: FULL INVOLUTES
The procedure is as follows (Fig. 5.4):

1 Draw the base circle.
2 Divide into twelve equal parts.
3 Determine the circumference of the base circle and divide into twelve.
4 Draw tangents from each circle division.
5 Mark off increasing circumferential distances along each succeeding tangent.
6 Sketch the involute curve between the tangent markings, e.g.

Tangent length XY = Arc length OY

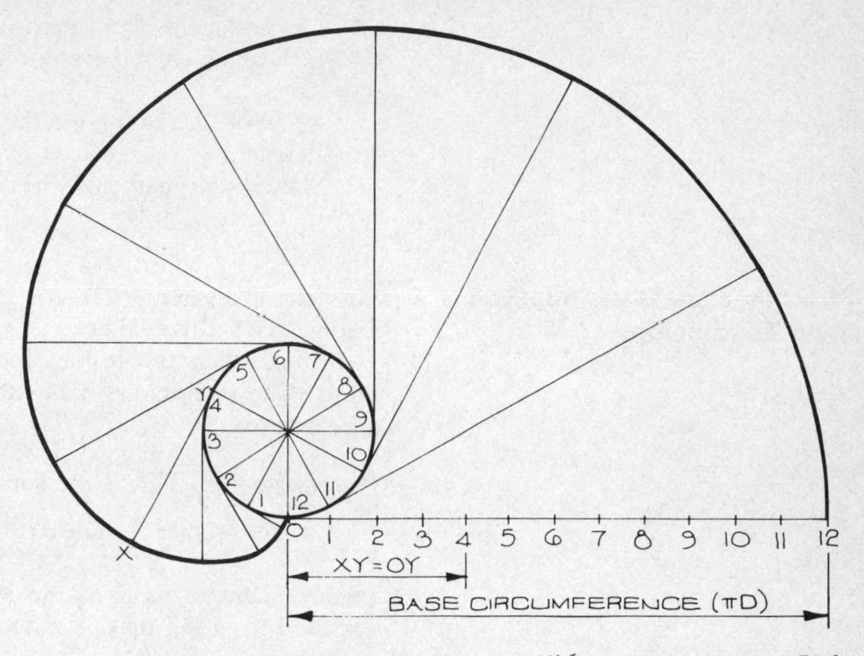

Fig. 5.4 Full involute

GEOMETRIC CONSTRUCTION: PART INVOLUTE (Fig. 5.5) The profile of a gear tooth is of course only a small part of the complete involute. To construct an accurate part-involute, the following procedure may be adopted:

1 Draw the base circle arc.

2 Mark off a number of equal small angles (say 10°).

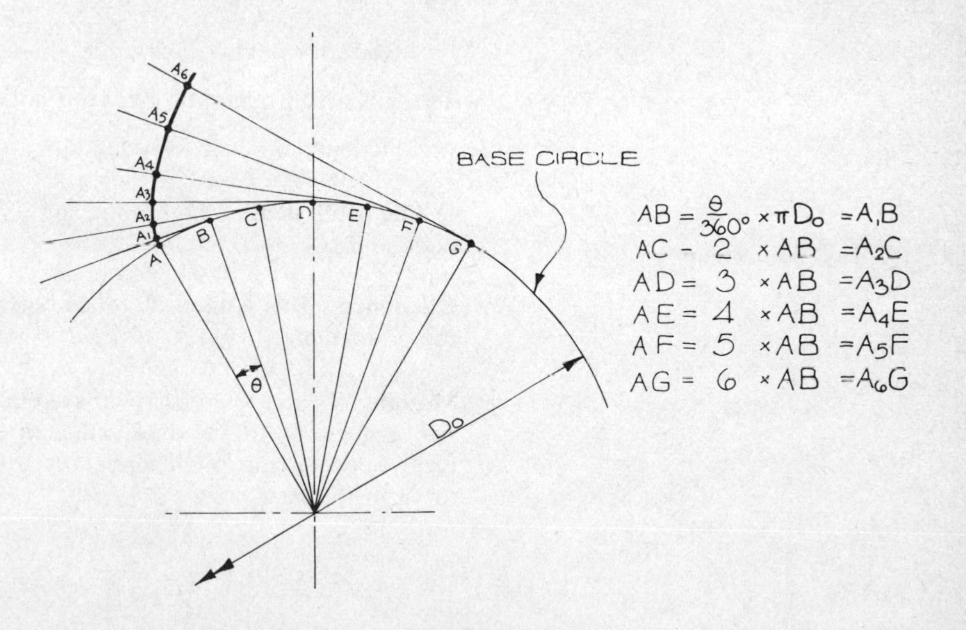

Fig. 5.5 Part involute

3 Draw tangents from each angular division.
4 Determine circumferential arc distances of angular divisions.
5 Mark off increasing arc distances along each succeeding tangent.
6 Sketch the part-involute curve between the tangent markings.

5.4 Involute Gear Proportions and Terminology

Pitch circle diameter (D for the gear wheel or d for the pinion) This is the diameter of the circle about which the two gears make their full rolling action with the absence of any sliding. The pitch circle diameters also affect the centre distance between gears.

Centre distance This is the sum of the pitch circle radii, i.e.

$$\text{Centre distance} = \tfrac{1}{2}(D + d)$$

Outside diameter This is the largest diameter of each gear measured from the tip of the tooth.

Root diameter This is the diameter of full tooth depth, i.e. the diameter which contains the bottoms of the tooth spaces.

Addendum A This is the height of the tooth above the pitch circle, i.e.

$$\text{Outside diameter} = \text{Pitch circle diameter} + 2A$$

Dedendum De This is the depth of tooth beneath the pitch circle. i.e.

$$\text{Root diameter} = \text{Pitch circle diameter} - 2De$$

There is a constant ratio between addendum and dedendum:

$$\text{Dedendum} = 1.25 \times \text{Addendum}$$

Whole tooth depth This is the sum of the addendum and the dedendum.

Clearance This is the difference between the dedendum and the addendum.

Module M In the metric gear system, the indication of tooth size is given by the module value. In effect the module is the ratio between the pitch circle diameter and the number of teeth in the gear, i.e.

$$M = \frac{D}{N}$$

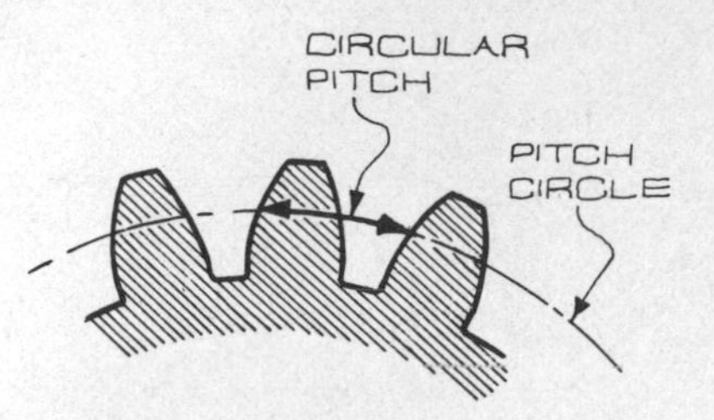

Fig. 5.6 Circular pitch

where M = module
D = pitch circle diameter
N = number of teeth.
This ratio is equivalent to the value of the addendum, i.e.

Module = Addendum

Diametral pitch *DP* This ratio between number of teeth and pitch circle diameter is extensively used in the Imperial (inch) unit gear system.

$$DP = \frac{N}{D}$$

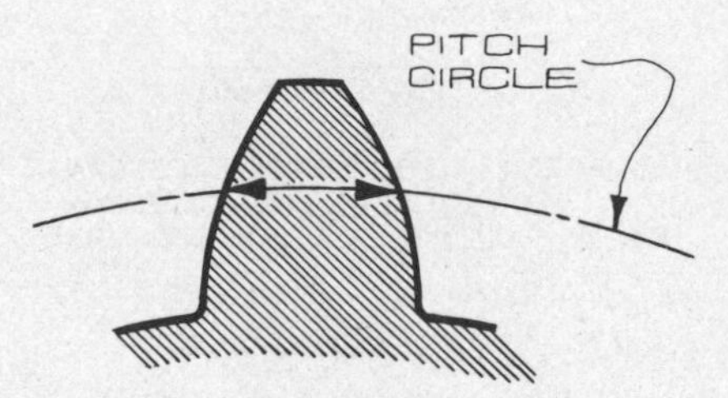

Fig. 5.7 Circular tooth thickness

i.e. the diametral pitch is the inverse of the module.

$$DP = \frac{1}{M} = \frac{1}{A}$$

Circular pitch *C* This is the circumferential distance along the pitch circle between corresponding points on the faces of adjacent teeth (Fig. 5.6).

$$\text{Circular pitch } C = \frac{\text{Pitch circle circumference}}{\text{Number of teeth}}$$

$$C = \frac{\pi D}{N} = \pi M$$

where M = module.

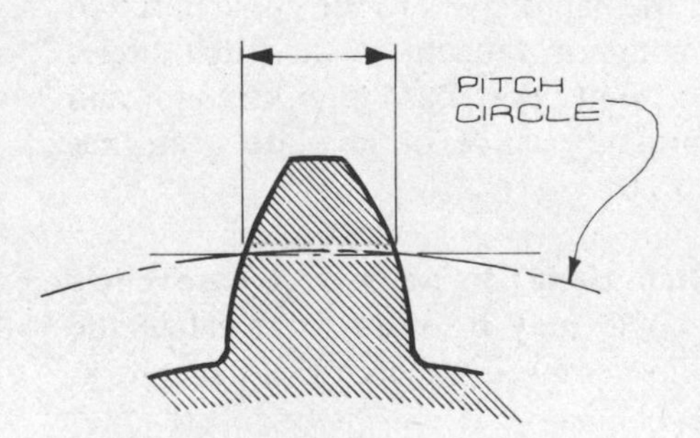

Fig. 5.8 Chordal tooth thickness

Circular tooth thickness This is the thickness of the tooth measured along the circumference of the pitch circle (Fig. 5.7).

$$\text{Circular tooth thickness} = \frac{\text{Circular pitch}}{2}$$

Chordal tooth thickness This is the thickness of the tooth measured along the chord of the arc (Fig. 5.8). For gear construction exercises at this level, the chordal tooth thickness will be considered as approximately equal to the circular tooth thickness.

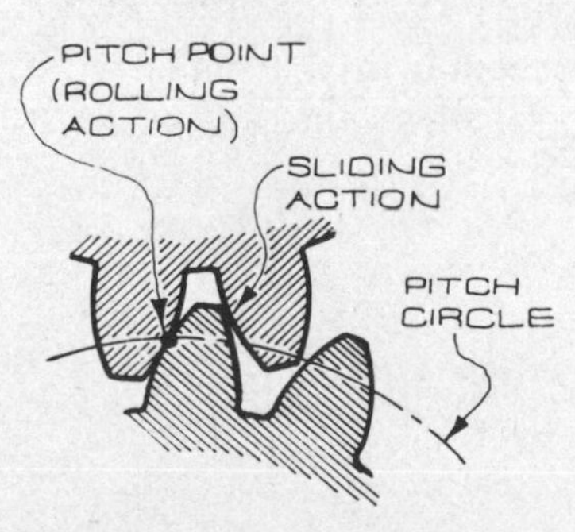

Fig. 5.9 Pitch point

Pitch point This is the point of contact between intersecting pitch circles (Fig. 5.9). When two gears are in contact at the pitch point, the tooth action is entirely rolling. Sliding of teeth takes place before and after contact at the pitch point.

Base circle As previously described in section 5.2 the base circle is the circle about which the involute curve is constructed. The diameter of the base circle is called the *base circle diameter* D_0.

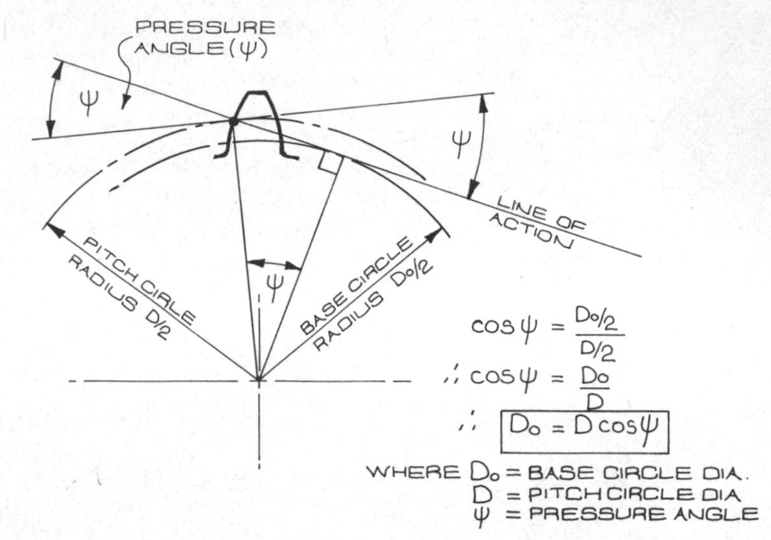

Fig. 5.10

Line of action This is the line along which contact between teeth of meshing gears takes place. The line of action passes through the pitch point tangential to the two base circles.

Pressure angle ψ This is the name given to the angle between the line of action and a common tangent to the pitch circles passing through the pitch point. For reasons of strength and meshing efficiency, the pressure angle of involute gears has now been standardized to 20°.

Relationship between pitch circle diameter and base circle diameter The pressure angle may be used to calculate the base circle diameter for any required pitch circle diameter, as shown in Fig. 5.10.

5.5 Gear Speed Ratios

The ratio of speeds between the driving pinion and driven gear wheel is in proportion to the number of teeth in each gear.

$$\frac{\text{Driven rev/min}}{\text{Driving rev/min}} = \frac{\text{Number of teeth in driving pinion}}{\text{Number of teeth in driven gear}}$$

or

$$\frac{\omega_2}{\omega_1} = \frac{n}{N}$$

where ω_2 = driven rev/min
ω_1 = driving rev/min
N = gear wheel teeth
n = pinion teeth.

5.6 Gear Example 1

A gear drive comprising two involute gears is to have a speed reduction of 1.5:1. The driving pinion has 20 teeth and revolves at 120 rev/min. If the gears are to have a module of 2 mm, determine

1. The speed of the driven gear wheel.
2. The number of teeth in the driven gear wheel.
3. The pitch circle diameter of each gear and their centre distance.
4. The outside diameter of each gear.
5. The root circle diameter of each gear.
6. The circular pitch.

Solution

1 Speed reduction = 1.5:1

$$\text{Driven gear wheel speed} = \frac{\text{Driving speed}}{1.5} = \frac{120}{1.5} = 80 \text{ rev/min} \quad (\textit{Ans})$$

2 $\dfrac{\omega_2}{\omega_1} = \dfrac{n}{N}$ i.e. $\dfrac{80}{120} = \dfrac{20}{N}$

$$N = \frac{120 \times 20}{80} = 30$$

Number of teeth in driven gear wheel = 30 (*Ans*)

3 For driven gear wheel

$$m = \frac{D}{N}$$

$$2 = \frac{D}{30}$$

$$D = 60 \text{ mm}$$

Pitch circle diameter of gear wheel = 60 mm (*Ans*)

For driving pinion

$$M = \frac{d}{n}$$

$$2 = \frac{d}{20}$$

$$d = 40 \text{ mm}$$

Pitch circle diameter of pinion = 40 mm (*Ans*)

$$\text{Centre distance} = \frac{D}{2} + \frac{d}{2} = \frac{60}{2} + \frac{40}{2} = 30 + 20 = 50 \text{ mm} \quad (\textit{Ans})$$

4 Addendum $A = \text{Module} = 2\ \text{mm}$
For driven gear wheel

$$\text{Outside diameter} = D + 2A = 60 + (2 \times 2) = 64\ \text{mm} \quad (Ans)$$

For driving pinion

$$\text{Outside diameter} = d + 2A = 40 + (2 \times 2) = 44\ \text{mm} \quad (Ans)$$

5 Dedendum $De = 1.25 \times \text{addendum} = 1.25 \times 2 = 2.5\ \text{mm}$
For driven gear wheel

$$\text{Root diameter} = D - 2De = 60 - (2 \times 2.5) = 55\ \text{mm} \quad (Ans)$$

For driving pinion

$$\text{Root diameter} = d - 2De = 40 - (2 \times 2.5) = 35\ \text{mm} \quad (Ans)$$

6 Circular pitch $C = \pi M = \pi \times 2 = 6.283\ \text{mm}$ (*Ans*)

5.7 Gear Example 2

The centre distance of two involute gears is 70 mm and the speed reduction is to be 2.5:1. If the module of the gears is 5 and the pressure angle is 20°, calculate

1 The pitch circle diameter of each gear.
2 The number of teeth in each gear.
3 The addendum.
4 The dedendum.
5 The total tooth depth.
6 The circular tooth thickness.
7 The base circle diameter of each gear.

Solution

1 Centre distance between gears is the sum of pitch circle radii with a 2.5:1 speed reduction. The driven gear wheel pitch circle radius is 2.5 times greater than that of the driving pinion.

If pinion pitch circle radius $= r$, then

$$\text{Gear wheel pitch circle radius} = 2.5r$$
$$r + 2.5r = \text{centre distance}$$

Therefore

$$r + 2.5r = 70$$
$$3.5r = 70$$
$$r = 20\ \text{mm}$$

Pitch circle diameter of driving pinion $= 2 \times 20 = 40$ mm
Pitch circle radius of driven gear wheel $= 2.5 \times 20 = 50$ mm
Pitch circle diameter of driven gear wheel $= 2 \times 50 = 100$ mm

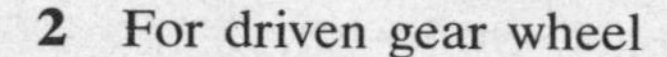

2 For driven gear wheel

$$M = \frac{D}{N}$$

$$5 = \frac{100}{N}$$

$$N = \frac{100}{5} = 20$$

Number of teeth in driven gear wheel = 20 (*Ans*)
For driving pinion

$$m = \frac{d}{n}$$

$$5 = \frac{40}{n}$$

$$n = \frac{40}{5} = 8$$

Number of teeth in driving pinion = 8 (*Ans*)
3 Addendum A = Module $M = 5$ mm (*Ans*)
4 Dedendum $De = 1.25 \times \text{addendum} = 1.25 \times 5 = 6.25$ mm (*Ans*)
5 Total tooth depth = Addendum + Dedendum

$$= 5 + 6.25 = 11.25 \text{ mm} \quad (Ans)$$

6 Circular tooth thickness $= \dfrac{\text{Circular pitch } C}{2}$

$$C = \pi M = \pi \times 5 = 15.7 \text{ mm}$$

so that

$$\text{Circular tooth thickness} = \frac{15.7}{2} = 7.85 \text{ mm}$$

7 For driven gear wheel

$$D_0 = D \cos \psi = 100 \times \cos 20° = 93.97 \text{ mm}$$

Base circle diameter of driven gear wheel = 93.97 mm (*Ans*)
For driving pinion

$$d_0 = d \cos \psi = 40 \times \cos 20° = 37.59 \text{ mm}$$

Base circle diameter of driving pinion = 37.59 mm (*Ans*)

5.8 Conjugate Tooth Action

An important requirement for rotary meshing action is that uniform tangential motion, transmitted at the driving pinion pitch circle, is maintained as uniform motion when transferred to the driven gear wheel. This condition would theoretically be satisfied by a pure rolling contact action between two cylindrical surfaces subjected to zero slip. Meshing teeth conforming ideally to this condition are called Conjugate Teeth.

The similarity of involute tooth action with that of pure conjugate teeth has been the most important factor in the evolution of the involute curve of the basic gear profile.

5.9 Construction of Involute Gear Teeth

Involute gear teeth are best drawn on tracing paper. The following example describes the recommended procedure.

Example

An involute gear system requires a 1:1 speed ratio and 20 teeth per gear. If the module is 10 mm and the pressure angle is 20°, draw five meshing teeth of the system.

Procedure (Refer to Fig. 5.11)

1. Determine the centre distance and draw in the pitch circles on a piece of tracing paper.
2. Mark the pitch point and draw the common tangent.
3. Draw the line of action for 20° pressure angle.
4. Calculate the base circle diameter and draw in the base circles (check that they are tangential to the line of action).
5. Calculate the addendum and draw the outside diameters.
6. Calculate the dedendum and draw the root circles.
7. Calculate the circular tooth thickness and step off several compass settings from the pitch point around each pitch circle.
8. Draw a part-involute to one base circle at a remote point from the pitch point using the procedure described in section 5.3.
9. Reproduce the involute form on a small piece of tracing paper.
10. Trace the involute at each tooth-thickness marking and complete the tooth forms.

Calculations

$$M = \frac{D}{N}$$

$$10 = \frac{D}{20}$$

$$D = 10 \times 20 = 200 \text{ mm}$$

Pitch circle radius = 100 mm
Centre distance = 100 + 100 = 200 mm

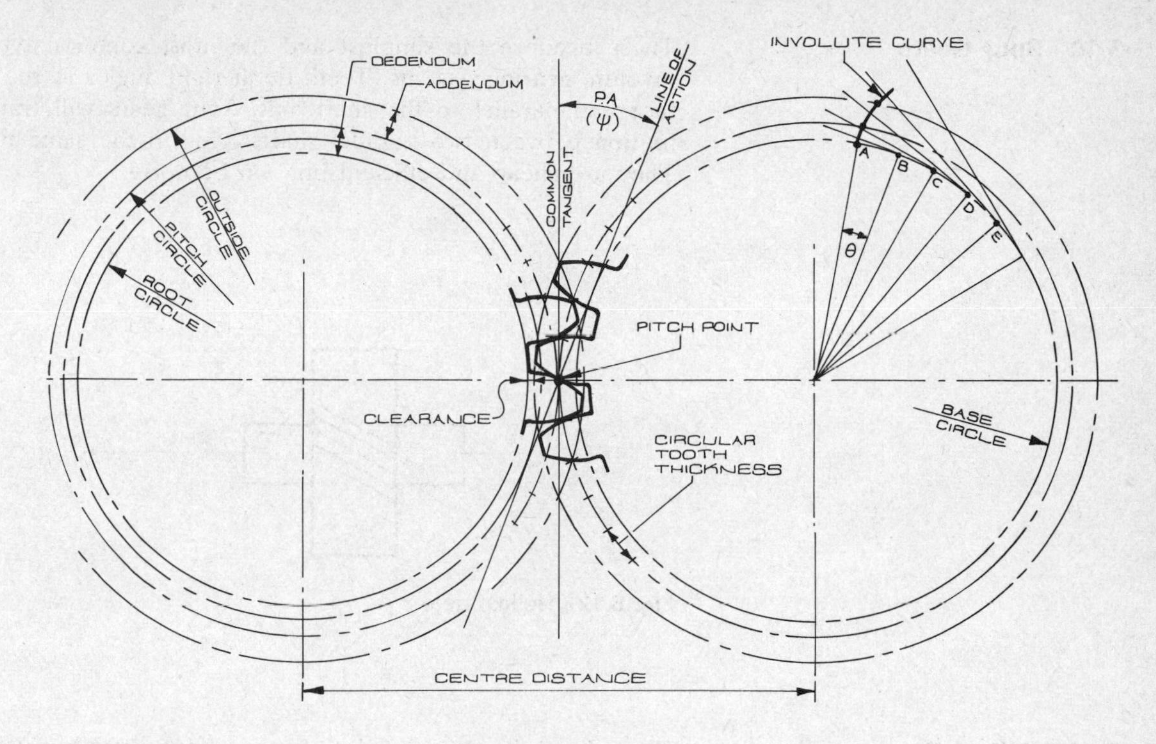

Fig. 5.11 Construction of involute gear teeth

$D_0 = D \cos \psi = 200 \times \cos 20° = 187.9$ mm

Base circle diameter = 187.9 mm

Radius of base circle $= \dfrac{187.9}{2} = 93.95$ mm

Addendum = Module = 10 mm
Outside radius = 100 + 10 = 110 mm
Dedendum = 1.25 × addendum = 1.25 × 10 = 12.5 mm
Root radius = 100 − 12.5 = 87.5 mm
Circular pitch $C = \pi M = \pi \times 10 = 31.42$ mm

Circular tooth thickness $= \dfrac{31.42}{2} = 15.71$ mm

Arc/tangent lengths on involute (for $\theta = 10°$):

$$AB = \frac{\theta}{360°} \times \pi D_0 = \frac{10}{360} \times \pi \times 187.9 = 16.4 \text{ mm}$$

$$AC = 2 \times AB = 2 \times 16.4 = 32.8 \text{ mm}$$

$$AD = 3 \times AB = 3 \times 16.4 = 49.2 \text{ mm}$$

$$AE = 4 \times AB = 4 \times 16.4 = 65.6 \text{ mm}$$

The involute gear tooth may take on many different forms. Some of the common types will now be discussed. Illustrations of these gear forms are shown in Chart 2 on page 139.

5.10 Spur Gears

These produce the simplest and the most common type of involute gearing systems. Teeth lie at right angles to the gear face and parallel to the shaft axis. Spur gears will transmit motion between two parallel shafts lying in the same plane. They are cheap and efficient but can be noisy.

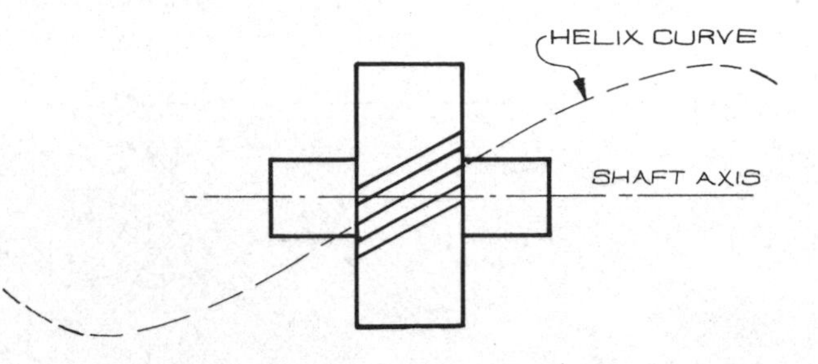

Fig. 5.12 Helical gear

5.11 Helical Gears

These have the same involute form as spur gears but, instead of the teeth lying parallel to the axis of the shaft, they form part of a helical curve (Fig. 5.12).

This type of arrangement produces more-gradual engagement and disengagement of teeth than that of spur gears. Helical gears thus produce a smoother drive, less vibration and noise, and a more evenly-spread tooth load. One serious disadvantage is the creation of an axial thrust load due to the tooth helix. The thrust will be transmitted to the shaft bearings and thus bearing selection is an important consideration for a helical drive.

5.12 Double Helical Gears

These may be employed to eliminate the helical thrust load previously described. One half of the face width has a right-hand helix and the other half has a left-hand helix (Fig. 5.13). This produces opposed thrust loads of equal magnitude which thus cancel each other out.

5.13 Bevel Gears

These are used to transmit motion between two shafts in the same plane but at right angles to each other. The involute teeth lie along part of a cone whose apex, if produced, would lie at the intersection of the shaft axes. The tooth thickness thus diminishes as the centre is approached.

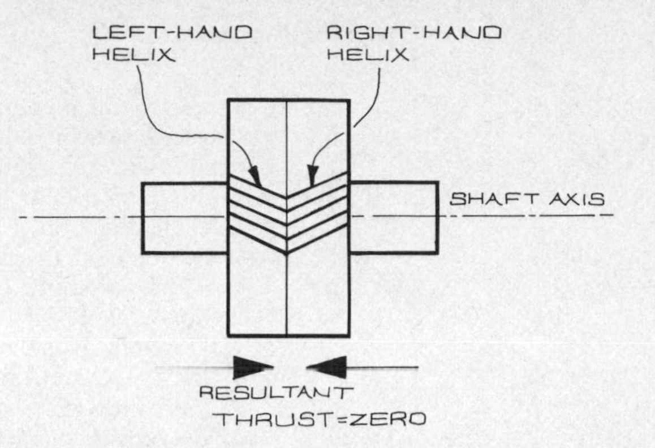

Fig. 5.13 Double helical gear

5.14 Spiral Bevel Gears

These may be considered as the equivalent of helical gears for a right-angle drive. The involute teeth lie along a helix formed around a cone (i.e. a conical spiral). Spiral bevel gears thus have the same smooth-running advantages as helical gears.

It should be noted that both forms of bevel gear will produce thrust loading which must be considered during choice of bearings.

5.15 Worm and Wheel

These produce a right-angled drive for shafts which are not in line. The worm is really a screw-thread with an involute form, which engages in teeth on the wheel.

A worm and wheel has the principal advantage of producing a very large reduction from a single reduction system. The resulting saving of materials makes this system cheap to produce in comparison with other drives. The chief disadvantage is that the tooth action is entirely sliding. The worm and wheel is thus the least efficient of gear systems, and most prone to tooth wear.

5.16 Rack and Pinion

This type of system is used to transmit rotary motion to straight-line motion, or vice versa. This pinion is standard involute spur gear, whilst the rack may be considered as the straight-line equivalent of a spur gear. Straight-line gearing is created when the pitch circle and base circle stretch to infinity.

Problems

5.1 Construct a full involute to a base circle of diameter 80 mm.

5.2 Construct a part-involute to a 70° sector of a base circle of radius 130 mm. (Use 10° angular spacings.)

5.3 A 20-tooth spur gear has a module of 4 mm. Calculate: *a*) the pitch circle diameter, *b*) the addendum, *c*) the dedendum, *d*) the root diameter, *e*) the outside diameter, *f*) the circular pitch.

5.4 A gear drive comprising two involute gears is to have a speed reduction of 2:1. The driving pinion has 30 teeth and revolves at 150 rev/min. If the gears have a module of 5 mm, determine
- *a*) the speed and number of teeth of the driven gear
- *b*) the centre distance
- *c*) the outside diameter and root circle diameter of each gear.

5.5 Define the terms: base circle, pitch point, line of action, pressure angle.

5.6 A spur gear has a pitch circle diameter of 160 mm and pressure angle of 20°. Calculate the base diameter and check the result with a full-size drawing.

5.7 The centre distance of two involute spur gears is 275 mm, and the speed reduction is 1.75:1. If the module of the gears is 2 mm and the pressure angle is 20° determine

The pitch circle diameter of each gear.
The number of teeth in each gear.
The addendum.
The dedendum.
The total depth of tooth.
The outside diameter of each gear.
The root diameter of each gear.
The base circle diameter of each gear.
The circular pitch.
The circular tooth thickness.

5.8 An involute gear system requires a 1:1 speed ratio and 18 teeth per gear. If the module is 12 mm and the pressure angle is 20°, draw five meshing teeth of the system.

5.9 Describe the characteristics of helical gears and explain why they are sometimes preferred to spur gears in mechanical drivers.

5.10 Helical gears have one physical disadvantage. Name this disadvantage and explain how it may be overcome.

5.11 The initial cost of a worm and wheel system is very low for the ratio achieved, but this gear arrangement may prove more expensive than other types in the long term. Explain.

5.12 Referring to Chart 2 on page 139, fill in the blank spaces against each gear-form shown, using brief notes regarding general description and application.

6 Limits, Fits, and Geometric Tolerancing

6.1 Toleranced Dimensions

No component may be produced exactly to the required size. Therefore if efficient assembly, location and operation are to be attained, all dimensions will be required to lie within a particular tolerance range. The extent of this range will depend on the design application. If accuracy is not critical, a general tolerance range will often be quoted on a company's standard drawing sheet, e.g. "All dimensions to be correct within ±0.1 mm unless otherwise stated".

A dimension with a particular tolerance range may be shown on a drawing in several forms. For example, in the case of Fig. 6.1, 40 mm is called the **nominal size**, i.e. the ideal size required.

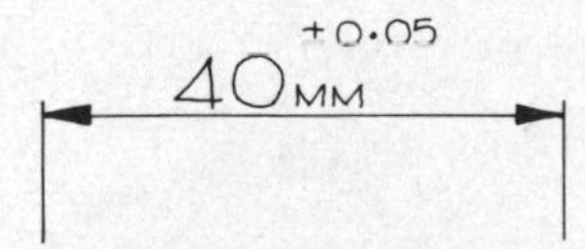

Fig. 6.1 Nominal size

The **upper limit** is the largest size allowed, i.e.

$$\text{Upper limit} = 40 + 0.05 = 40.05 \text{ mm}$$

The **lower limit** is the smallest size allowed, i.e.

$$\text{Lower limit} = 40 - 0.05 = 39.95 \text{ mm}$$

The **tolerance** is defined as the difference between the upper limit and lower limit, i.e.

$$\text{Tolerance} = 40.05 - 39.95 = 0.1 \text{ mm}$$

An alternative display of this dimension is to quote the upper and lower limits on the drawing, as shown in Fig. 6.2.

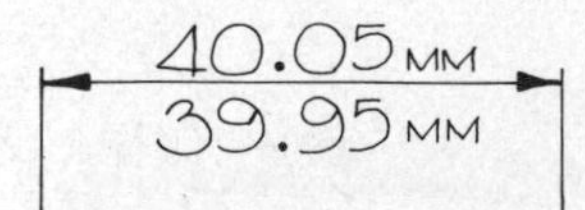

Fig. 6.2 Upper and lower limits

6.2 Types of Fit

A fit is defined as a working condition between a mating shaft and hole.

Clearance fit This gives a condition in which the shaft is smaller than the hole under all extremes, i.e. the upper limit of the shaft is smaller than the lower limit of the hole. Clearance fits are most commonly used where rotational or lateral movement occurs between male and female components, such as a spindle inside a plain bearing.

Interference fit This gives a condition in which the hole is smaller than the shaft under all extremes, i.e. the upper limit of the hole is smaller than the lower limit of the shaft. Interference fits are often used where a rigid fixing is required, such as a gear or collar bore on a shaft.

Transition fit This may provide either a clearance or an interference at the extremes of fit. One of the most common applications of transition fits is in keys and keyways.

Determining the type of fit This may be done by comparing the extremes of male and female sizes.

The *largest shaft* is compared with the *smallest hole.*
The *largest hole* is compared with the *smallest shaft.*
An example is shown in Fig. 6.3.

Largest hole	= 50.05 mm
Smallest shaft	= 49.99 mm
Difference	= 0.06 mm (clearance)
Largest shaft	= 50.02 mm
Smallest hole	= 50.03 mm
Difference	= 0.01 mm (clearance)
Extremes:	0.06 mm clearance
	0.01 mm clearance

Thus, the condition is a clearance fit.

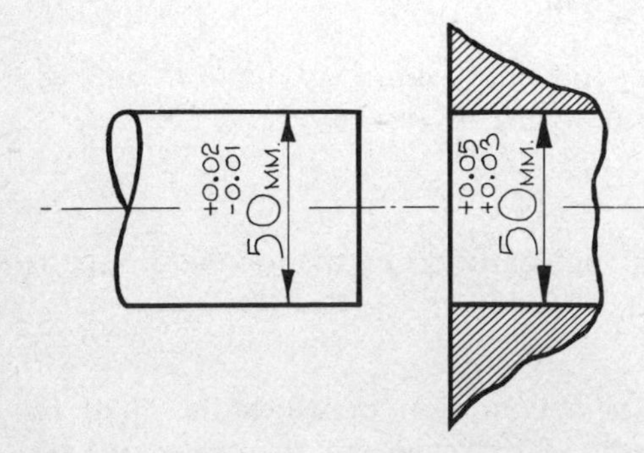

Fig. 6.3 Shaft and hole fit

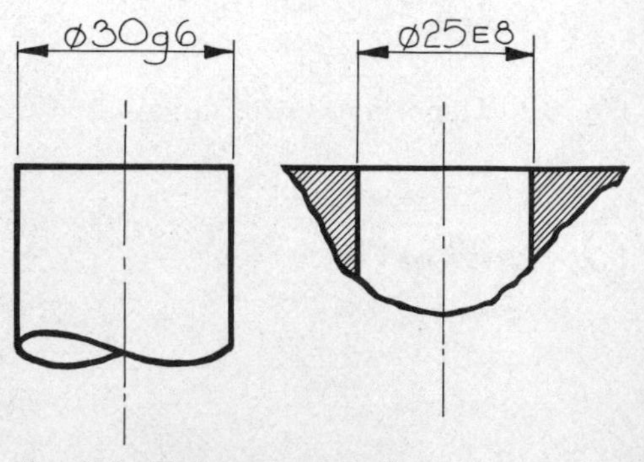

Fig. 6.4 Tolerance notation

6.3 British Standard Limits and Fits (BS 4500)

This standard governs the range of limits and fits used for holes and shafts in industry.

There are twenty seven available **fundamental variations** from the nominal size.

Deviations for holes are denoted by capital letters (A–Z).

Deviations for shafts are denoted by small letters (a–z).

For each size variation there are eighteen possible **tolerance bands**. These are denoted by numbers (01, 0, 1, 2, 3, . . . , 16).

A toleranced dimension conforming to BS 4500 will thus be denoted by a letter and a number against the nominal size, as shown in Fig. 6.4.

Some of the more common combinations of shaft and hole are listed in Chart 3 on pages 140–141.

6.4 Geometric Tolerancing

A dimension tolerance restricts a size to certain limits. Similarly a **geometric tolerance** restricts a component to limits of shape. For example, a shaft may be correct to a certain size of diameter but may not be a correct circular shape. Geometric tolerances are listed in Part 3 of BS 308 under various symbols for tolerance of shape. The more common shapes and methods of drawing representation are depicted on pages 72 and 73.

Straightness A surface or axis may be deemed to be adequately straight if it lies between two parallel straight lines of specified distance apart.

A straightness tolerance limits the amount of "waviness" of a surface in two dimensions.

Flatness A surface is adequately flat if it lies between two parallel planes of specified distance apart.

A flatness tolerance limits the amount of "bumpiness" of a surface in three dimensions.

Roundness A surface is adequately round if it lies between two concentric circles of specified distance apart.

A roundness tolerance limits the amount of ovality in a circular shape.

Cylindricity A surface is adequately cylindrical if it lies between two concentric cylinders of specified distance apart.

A cylindricity tolerance limits the amount of ovality on a cylinder's cross-section and also the "bumpiness" along its length.

Parallelism A surface is adequately parallel to a datum surface or axis, if it lies within two planes of specified distance apart and which are both parallel to the datum. An axis is adequately parallel to a datum surface or axis if it lies within a cylinder of specified diameter whose axis is parallel to the datum.

A parallelism tolerance limits the extent to which parallel surfaces are out of true.

Squareness A surface is adequately square to a datum surface or axis if it lies within two parallel planes of specified distance apart, both of which are square to the datum. An axis is adequately square to a datum surface or axis if it lies within a cylinder of specified diameter whose axis is square to the datum.

A squareness tolerance limits the extent to which perpendicular sufaces or axes may be out of true.

Angularity A surface is adequately correct to a stated angle from a datum surface if it lies within two parallel planes of specified distance apart, which are true to the required angle from the datum.

If one surface must lie at a stated angle to another, an angularity tolerance limits the extent to which two surfaces may be out of true.

Concentricity A cylinder is adequately concentric to a datum cylinder if its axis lies within a cylinder of specified diameter whose axis is in line with the axis of the datum cylinder.

Symmetry A width is adequately symmetrical about a datum width if its central axis lies within two parallel planes of specified distance apart and which are symmetrical about the central axis of the datum width.

A symmetry tolerance limits the extent to which two symmetrical widths may be out of true.

True position An axis is close enough to the true stated position if it lies within a cylinder of specified diameter whose axis is in the true stated position.

A positional tolerance limits the extent to which an axis may deviate from a stated position in three dimensions.

Problems

6.1 A shaft is made to the dimension 40 mm ± 0.05. State the nominal dimension, the upper limit, the lower limit, and the tolerance.

6.2 For each of the dimensions shown below, state the upper limit, the lower limit, and the tolerance.

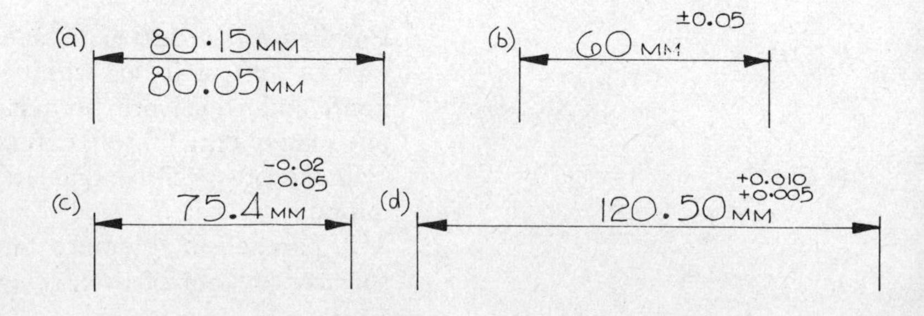

6.3 For each of the following combinations, calculate the extremes of fit. Hence, state whether each fit is a clearance, interference, or transition.

	Hole (mm)	Shaft (mm)		Hole (mm)	Shaft (mm)
(*a*)	80.030 80.000	80.021 80.002	(*e*)	450.063 450.000	450.045 450.005
(*b*)	250.115 250.000	249.000 249.785	(*f*)	55.046 55.000	54.970 54.940
(*c*)	30.025 30.000	29.991 29.975	(*g*)	100.035 100.000	99.988 99.966
(*d*)	150.040 150.000	150.068 150.043	(*h*)	200.046 200.000	200.079 200.060

6.4 Using the charts from BS 4500 provided on pages 140–141, determine the upper and lower limits, extremes of fit, and the type of fit for the following shaft/hole combinations:

a) Shaft ϕ15 mm p6, Hole ϕ15 mm H7.
b) Shaft ϕ190 mm k6, Hole ϕ190 mm H7.
c) Shaft ϕ70 mm e9, Hole ϕ70 mm H9.
d) Shaft ϕ480 mm n6, Hole ϕ180 mm H7.
e) Shaft ϕ20 mm f7, Hole ϕ20 mm H8.
f) Shaft ϕ35 mm s6, Hole ϕ35 mm H7.

6.5 Conforming to BS 308 Part 3, indicate the following geometric tolerances on the component shown in Fig. 6.5:

a) Squareness of ϕ22 axis with the base within 0.03 mm diameter cylinder.
b) Concentricity of ϕ22 with ϕ30 within 0.04 mm diameter cylinder.
c) Symmetry of 14 mm slot with 16 mm slot within two parallel planes 0.02 mm apart.
d) 70° chamfer to be true to this angle within two parallel planes 0.03 mm apart.
e) ϕ15 hole to be true to the position stated within 0.03 mm dia. cylinder.

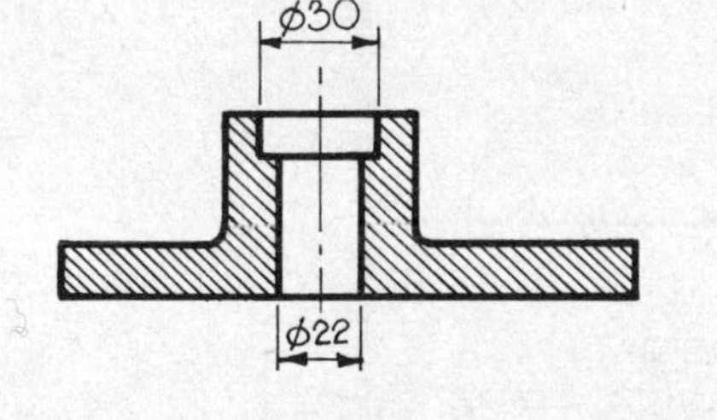

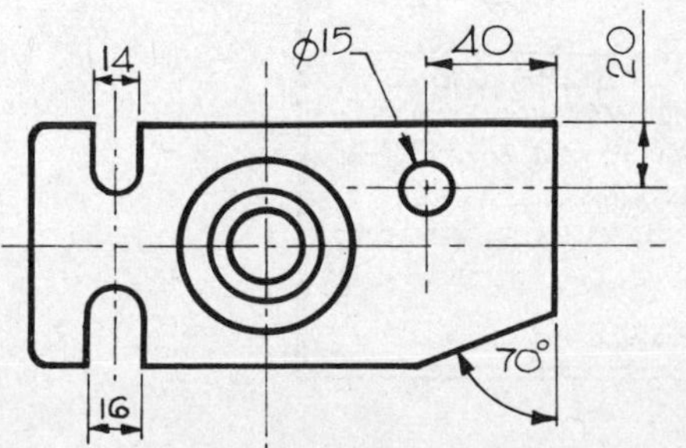

Fig. 6.5

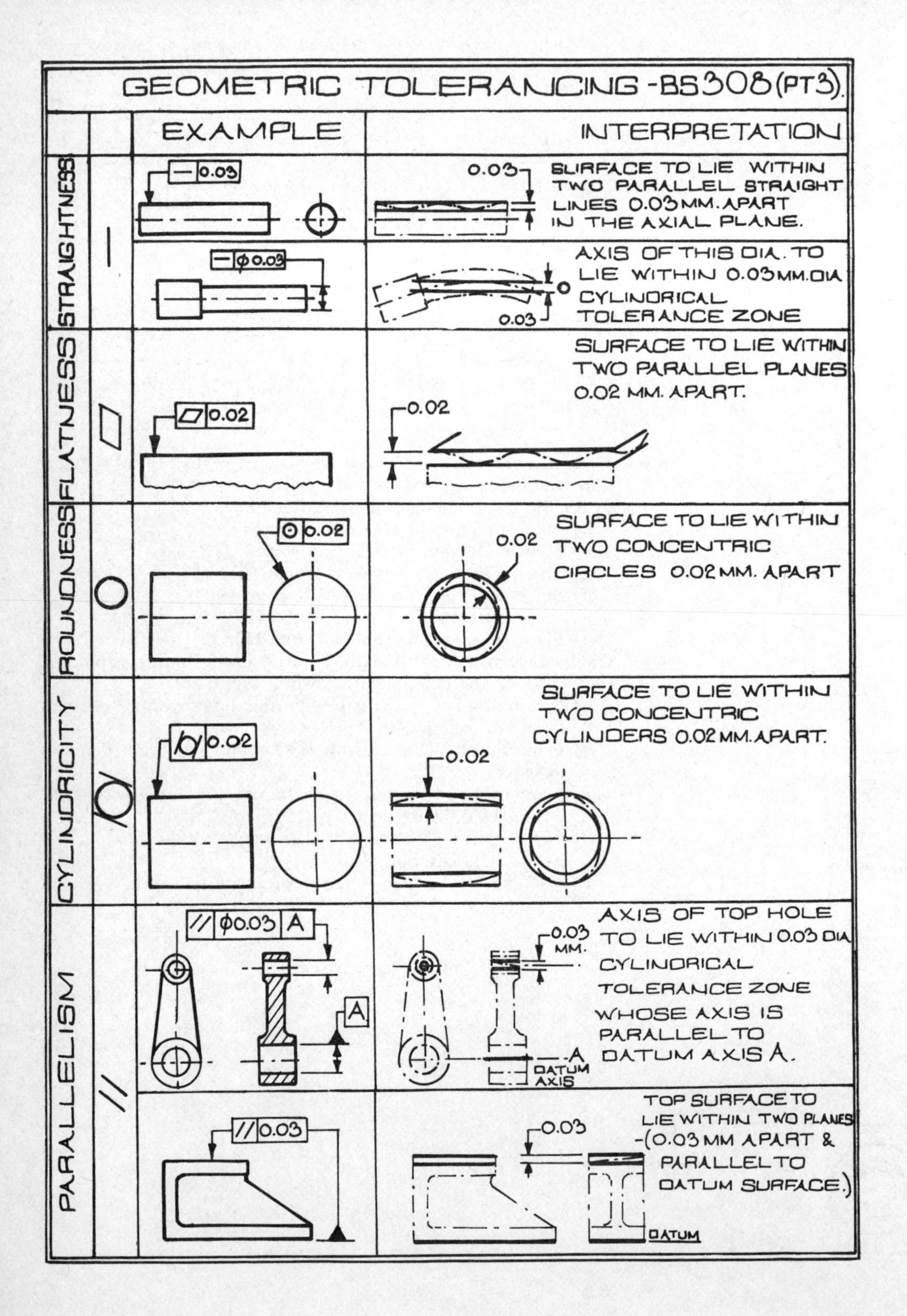
GEOMETRIC TOLERANCING - BS308(PT3).
EXAMPLE
INTERPRETATION
STRAIGHTNESS
0.03
0.03
SURFACE TO LIE WITHIN TWO PARALLEL STRAIGHT LINES 0.03 MM. APART IN THE AXIAL PLANE.
ϕ0.03
AXIS OF THIS DIA. TO LIE WITHIN 0.03 MM. DIA CYLINDRICAL TOLERANCE ZONE
0.03
FLATNESS
0.02
SURFACE TO LIE WITHIN TWO PARALLEL PLANES 0.02 MM. APART.
0.02
ROUNDNESS
0.02
0.02
SURFACE TO LIE WITHIN TWO CONCENTRIC CIRCLES 0.02 MM. APART
CYLINDRICITY
0.02
SURFACE TO LIE WITHIN TWO CONCENTRIC CYLINDERS 0.02 MM. APART.
0.02
PARALLELISM
// ϕ0.03 A
A
0.03 MM.
A
DATUM AXIS
AXIS OF TOP HOLE TO LIE WITHIN 0.03 DIA CYLINDRICAL TOLERANCE ZONE WHOSE AXIS IS PARALLEL TO DATUM AXIS A.
// 0.03
0.03
DATUM
TOP SURFACE TO LIE WITHIN TWO PLANES -(0.03 MM APART & PARALLEL TO DATUM SURFACE.)

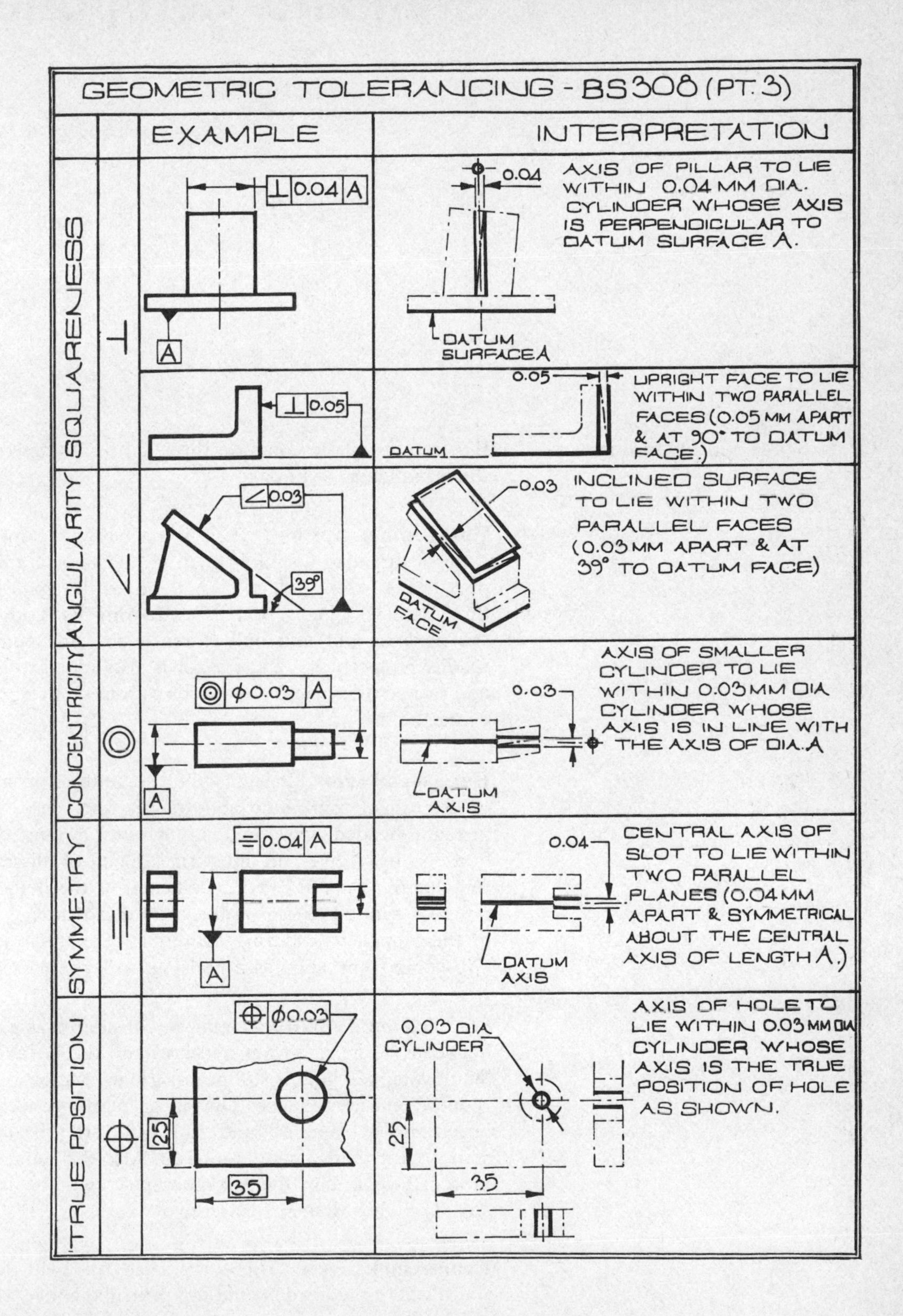
GEOMETRIC TOLERANCING - BS308 (PT. 3)
EXAMPLE
INTERPRETATION
SQUARENESS
⊥ 0.04 A
A
0.04
AXIS OF PILLAR TO LIE WITHIN 0.04 MM DIA. CYLINDER WHOSE AXIS IS PERPENDICULAR TO DATUM SURFACE A.
DATUM SURFACE A
⊥ 0.05
0.05
UPRIGHT FACE TO LIE WITHIN TWO PARALLEL FACES (0.05 MM APART & AT 90° TO DATUM FACE.)
DATUM
ANGULARITY
∠ 0.03
39°
0.03
INCLINED SURFACE TO LIE WITHIN TWO PARALLEL FACES (0.03 MM APART & AT 39° TO DATUM FACE)
DATUM FACE
CONCENTRICITY
◎ ϕ0.03 A
A
0.03
AXIS OF SMALLER CYLINDER TO LIE WITHIN 0.03 MM DIA CYLINDER WHOSE AXIS IS IN LINE WITH THE AXIS OF DIA. A
DATUM AXIS
SYMMETRY
⌯ 0.04 A
A
0.04
CENTRAL AXIS OF SLOT TO LIE WITHIN TWO PARALLEL PLANES (0.04 MM APART & SYMMETRICAL ABOUT THE CENTRAL AXIS OF LENGTH A.)
DATUM AXIS
TRUE POSITION
⌖ ϕ0.03
25
35
0.03 DIA CYLINDER
25
35
AXIS OF HOLE TO LIE WITHIN 0.03 MM DIA CYLINDER WHOSE AXIS IS THE TRUE POSITION OF HOLE AS SHOWN.

7 Standard Components

7.1 Threaded Fastenings

Illustrations of the common threaded fastenings discussed are shown in Chart 4 on page 142.

Hexagon nut and bolt This is probably the most common form of threaded fastening used in mechanical engineering. It provides a strong, rigid fixing and is the most efficient of threaded clamping actions. Considering its loading capacity, the hexagon nut and bolt is reasonably inexpensive, and is readily replacable. A flat washer is also added to assist seating and protect the component surface from damage due to tightening the nut.

Hexagon setscrew Inaccessibility sometimes prohibits the use of a hexagon nut. One solution in such cases is to use a hexagon-headed setscrew mating with a tapped hole. The hexagon head gives the most rigid fixing of all setscrews, but the clamping action is not as efficient as the hexagon nut and bolt. As with all setscrews the chief disadvantage is that wear of the tapped hole during maintenance may cause an expensive component, such as a housing, to be replaced.

Stud and nut This device may be considered as a compromise between the hexagon nut and bolt and the hexagon setscrew. While a tapped hole is still necessary, its rate of wear is greatly reduced in maintenance. During dismantling, wear is experienced only between nut and stud. If the stud thread wears, the stud is easily and cheaply replaced. Also the clamping action is more efficient than the hexagon setscrew. The initial cost is slightly higher than the setscrew.

Countersunk screw These are used for light-duty applications, where the head should not protrude above the surface of the fixing. Countersunk screws are cheap devices but are low in rigidity and sometimes subject to location difficulties.

Socket-head capscrew These perform a similar function to countersunk screws. They are more expensive, but give a stronger joint with better locking action and greater ease of location.

Round-headed screw This is a general-purpose screw which may be considered as the light-duty alternative to the hexagon setscrew.

Grub screw These are used almost exclusively to prevent axial or radial movement between female components and shafts. They are very cheap devices but should not be subjected to any mechanical loading. The shaft is usually "dimpled" to take the grubscrew.

7.2 Locking Devices

Illustrations of the locking devices discussed are shown in Chart 5 on page 143.

Nut and locknut This provides a good mechanical lock for a wide range of load capacities, and is used extensively throughout all branches of mechanical engineering.

Spring washer This cheap device relies on the principle of a compression spring. When fastened, the small spring provides a tension force which deters the threaded fastening from unfastening itself.

Star (shakeproof) washer This embeds itself into the metal surfaces and then relies on friction to prevent unfastening. This device is used mainly on light-duty fixings.

Tab washer This gives a good locking action for a wide range of loading capacities. Tab washers should be renewed after maintenance.

Castle nut and split pin This provides one of the most efficient locking actions and is used mainly on the higher duty applications. Although comparatively expensive, this device is essential for the safe operation of many transmission designs such as wheel axles.

Self-locking nut (nylon-insert) This gives a tight-fixing due to the bolt thread cutting its own path through the nylon insert. This type of nut should usually be replaced after maintenance.

Self-locking nut (all-metal) This is used for similar applications to the nylon-insert nut. It usually provides a locking action due to a cross-thread principle.

Wired bolts These provide a locking action almost as efficient as the castle nut and split pin and are slightly cheaper. They are used mainly where there is a risk of excessive vibration, e.g. in vehicle differential units.

7.3 Plain Bearings

A bearing may be defined as a component which gives radial or axial support to a rotating shaft.

A plain bearing is one in which circumferential sliding is allowed to take place between itself and the shaft. The resulting effects of friction must thus be accounted for. The required material properties for plain bearings are

1. Hardness.
2. Low coefficient of friction, or self-lubricating properties.
3. Good heat-dissipation properties.
4. Good strength in compression.

Typical materials rating high in these properties are phosphor bronze, grey cast iron, white metal (tin/lead base alloys), and certain grades of nylon.

Plain bearings are more often used for light-duty loading and slower speeds, or in machinery with short and infrequent usage. In most cases they are used in conjunction with a lubricant, although some plain bearings are completely self-lubricating. A typical example here is the development of oil-impregnated bushes made from porous sintered bronze. The oil inside the bush provides adequate lubricating and lasts the normal life of the bearing.

7.4 Anti-friction Bearings

Unlike plain bearings, anti-friction bearings dispense with the sliding effect by making use of rolling action between shaft and housing. The roller material must be hard and tough. For this reason a chromium alloy steel is usually employed. There are numerous types of anti-friction bearings, serving particular applications. Some of the more common types will now be discussed, with the corresponding illustrations shown in Chart 6 on page 144. Although general reference to loading capacities will be made, it should be realized that the selection of a bearing for loading capacity will depend not only upon radial and axial forces, but upon speed of rotation, lubrication, axial alignment, and frequency of use. In particular, the loading capacity of any bearing diminishes rapidly with increase of speed and thus a speed factor will always be applied to the loading calculation.

Needle roller bearing This is a bearing suitable for radial loads only. With small forces and good lubrication, high speeds are possible. The chief use of needle roller bearings is in providing rolling action inside a restricted space, such as exist in many types of oil pumps.

Rigid ball journal bearing This is the most commonly used type of anti-friction bearing. It will accommodate both radial loads and axial thrust loads for a wide capacity and speed range. It may also be used to locate a shaft in either axial direction.

Cylindrical roller bearings This may take very high loading for a wide speed range but will only accommodate radial loads. It will not take any axial thrust loading and will not locate a shaft since free axial movement is allowed between shaft and bearing. A particular application here is to allow for the expansion of a shaft at one end when subjected to heat. The other end is usually supported by a ball bearing to locate the shafts and take axial loading.

Thrust bearing This will take axial thrust loads only. It is most commonly used to support the gravity loads of heavy vertical shafts, such as in agitators and cooling fans, and is usually limited to the lower speed ranges.

Taper roller bearing This will accommodate both radial and axial thrust loading for medium and high loading. Its axial thrust loading capacity is far superior to the rigid ball bearing, especially at the higher speeds. Due to the nature of its design, the taper roller bearing will take axial thrust loads in one direction only, and therefore another opposed taper roller bearing will normally be fitted at the other end of the shaft. The bearing assembly is in fact held together by the axial thrust load and thus both bearings must be correctly adjusted to control end float.

Angular contact bearing This may be considered as the equivalent to the taper roller bearing for lighter loads. Again axial thrust is taken in one direction only and two opposed bearings are fitted while eliminating end float. As with the taper-roller this bearing will take radial loads in conjunction with axial thrust.

Self-aligning ball bearing One essential requirement of most bearings is that the shaft be accurately aligned with the housing for the load capacity to be maintained. When this cannot be guaranteed, the self-aligning ball bearing may be used. One particular application is that of long shafts which are liable to deflect and become misaligned. The angular displacement may be allowed for in the design of this bearing.

Self-aligning roller bearing This is very similar in principle and application to the self-aligning ball bearing, but in general the bearing will take higher radial loading.

7.5 Seals

A seal is a component part which prevents leakage of liquid (e.g. lubricant) around a moving item and also protects the mechanisms from ingress of dirt, etc. Illustrations of common types of seal are shown in Chart 7 on page 145. In choosing the type of seal required, the main considerations will be those of shaft speed, internal and external pressure, axial alignment and the type of liquid (e.g. type of lubricant) being sealed.

Felt washer This provides a cheap seal for grease lubrication. When fitted into its end cap, the washer becomes compact against the shaft and creates a seal. This device should be used only for slower speeds.

Garter spring seal This is probably the most common type of seal used in mechanical design. The garter spring forces the synthetic rubber lip against the shaft, via a coherent film of oil, to provide an efficient sealing mechanism against leakage. This type of seal may be used for a wide range of pressures and shaft speeds provided the shaft is machined to a good surface finish.

V-ring seal This has the advantage of fitting onto the shaft and seating between lip and housing. Thus, being made entirely of synthetic rubber, it will allow some axial misalignment and ovality of shaft. A good surface finish is not essential on the shaft but will be required at the rubbing surface of the lip.

Labyrinth seal This relies on the principle of a cover, fixed to the rotating shaft, and machined to a complex labyrinth contour on one side. This mates with a similar contour on the housing. If the gap between labyrinth faces is small and packed lightly with grease, an effective seal for oil or grease systems is achieved. Labyrinth seals find particular use where outside conditions, such as a dusty atmosphere, would cause abrasive damage to other types of seal.

O-ring This is merely a synthetic rubber-ring of circular cross-section. The O-ring does not strictly fit the required definition of a seal since it should not be used in conjunction with moving parts. However, this device is used extensively to prevent lubrication or hydraulic fluid leakage from covers and stationary shafts. The ring is normally housed in a square-section machined groove of specified width and depth.

Gland This device operates by the effect of compressing a packing material into a stuffing box. The shape of the gland bushes then ensures that the packing forces itself against the rotating shaft and creates a seal. The compression of the packing is usually achieved by tightening a gland cap into the stuffing box via threaded studs. A gap must be allowed between cap and housing to allow for adjustment during and after assembly. The packing usually consists of fibrous asbestos impregnated with graphite. Glands are commonly used to prevent liquid leakage from oil pumps and hydraulic valves.

Screwed gland This is basically a cheaper version of the conventional gland previously described. The packing here is compressed via a simple hexagon nut. This device is often used in small hydraulic valves.

Problems

7.1 Complete the assemblies shown in Fig. 7.1 using the correct threaded fastenings and washers shown in Charts 4 and 5 (pp. 142 and 143).

7.2 In Chart 4 (p. 142), fill in the blank spaces against each type of threaded fastening shown, using brief notes for description and application.

7.3 In Chart 5 (p. 143), fill in the blank spaces against each type of locking device shown, providing brief notes on description and application.

7.4 In Chart 6 (p. 144), fill in the blank spaces against each type of anti-friction bearing shown, giving brief notes on description and application.

7.5 In Chart 7 (p. 145), fill in the blank spaces against each type of seal or gland shown, giving brief notes on description and application.

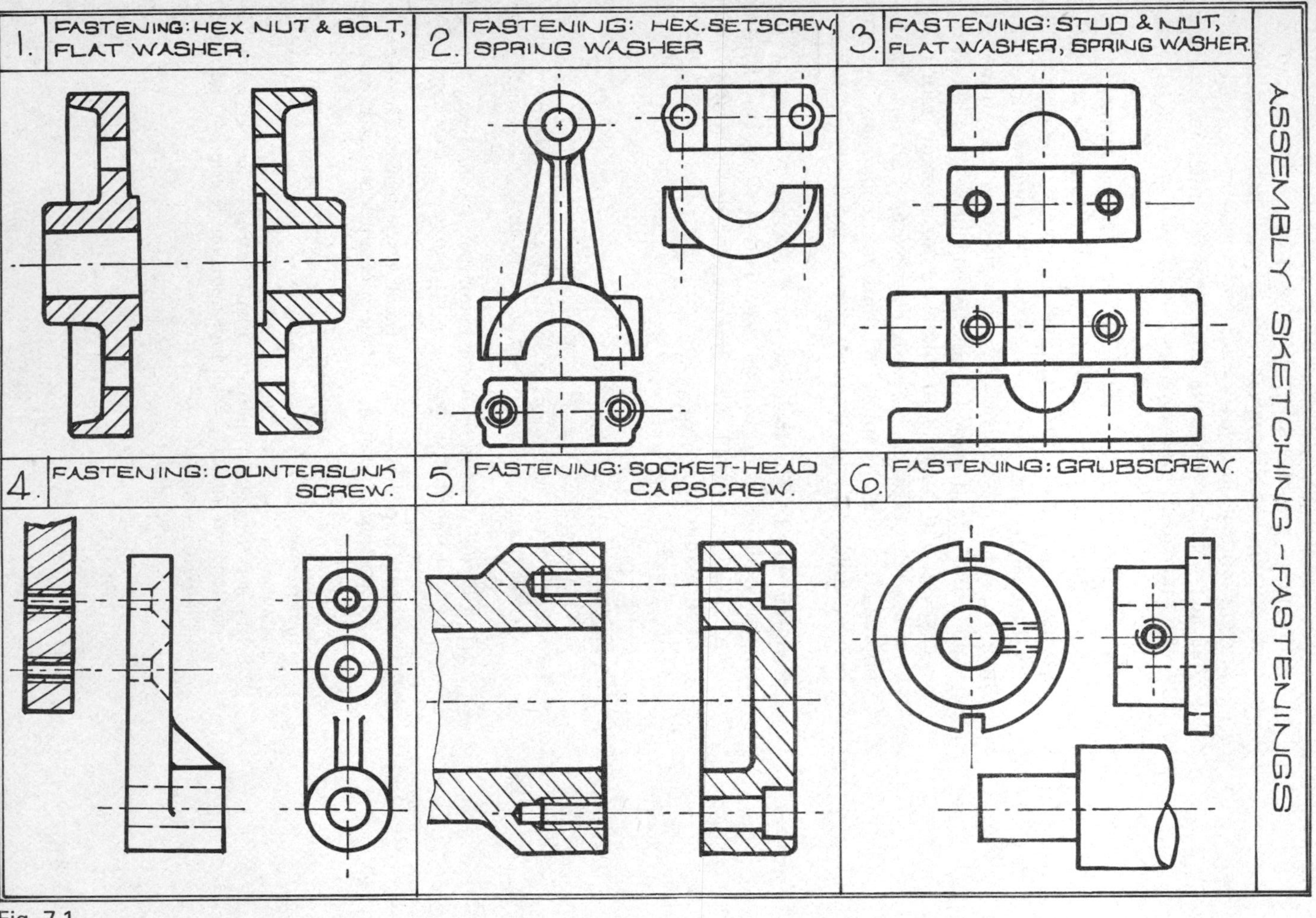
ASSEMBLY SKETCHING - FASTENINGS
1. FASTENING: HEX NUT & BOLT, FLAT WASHER.
2. FASTENING: HEX. SETSCREW, SPRING WASHER
3. FASTENING: STUD & NUT, FLAT WASHER, SPRING WASHER
4. FASTENING: COUNTERSUNK SCREW.
5. FASTENING: SOCKET-HEAD CAPSCREW.
6. FASTENING: GRUBSCREW.

Fig. 7.1

8 Systematic Design

8.1 Introduction

Previous chapters have discussed the physical, mechanical and chemical properties of materials. Consideration has also been given to methods of tolerancing drawings in order to simplify the means by which a component drawing can impart to the production departments the requirements of the designer. The foregoing information is part of the vast knowledge that the engineering designer needs in order to prepare a workable design.

In this section consideration will be given to the designer's art and how alternative design thinking can progress a design specification through to the final production scheme.

It is important that the designer is objective in thinking about design problems, drawing on basic knowledge at all times, and developing new ideas from the wealth of knowledge that is available. In the main, design of a new product is the re-shaping of existing ideas and turning them into a form suitable to fulfil the requirements of the new specification. The ability to innovate is one of the prime attributes that the designer must have. The ability to be creative and not just copy what has gone before will ensure that the product, whatever it is, becomes commercially viable, provided, of course, that careful thought has been given to the cost factor of the individual component parts.

In any job function it is evident that efficiency can only be achieved by applying logical steps. In engineering design a systematic approach is essential to achieve an efficient outcome to the design process.

8.2 Design Requirements

Before any design is commenced there must be a basic specification stating the objectives which must be met. These objec-

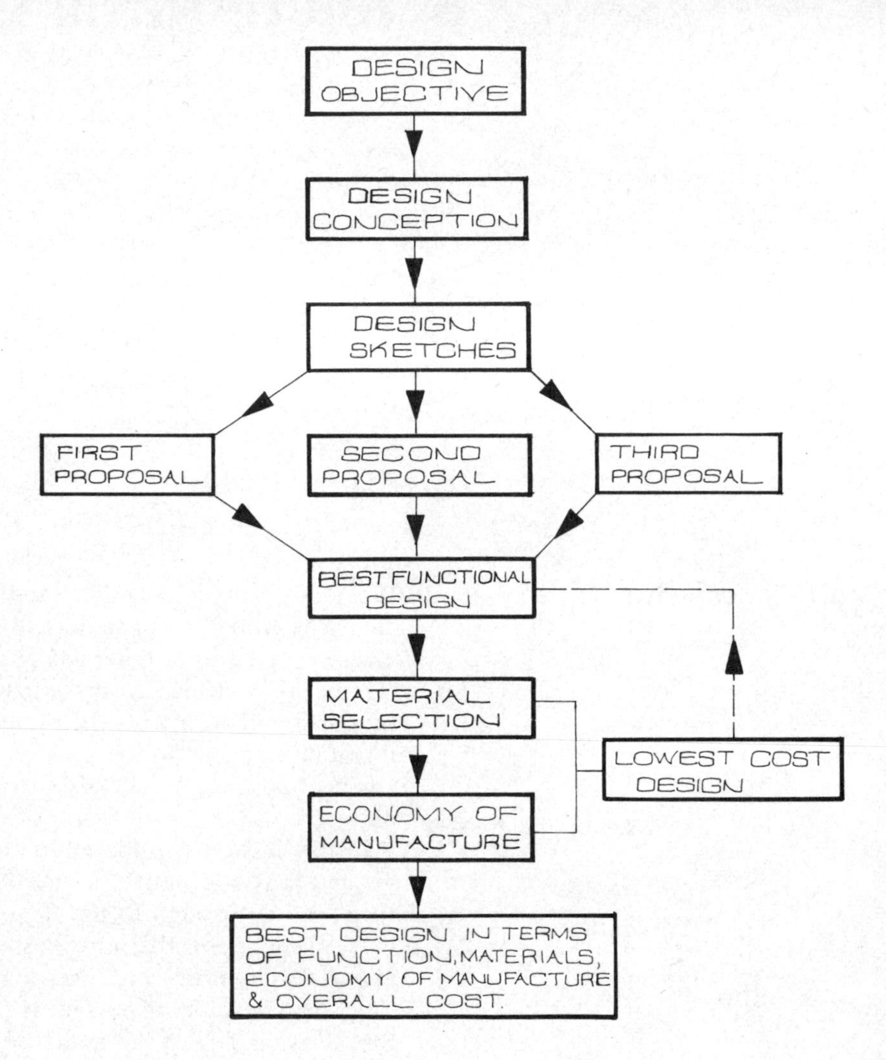

Fig. 8.1 The design process

tives may be set by the customer or be the result of the company wishing to improve its existing product range. The designer will be faced with the need to prepare sketches of designs which will satisfy the design specification.

It is advisable to consider more than one design appraisal, and to devise a method whereby a process of elimination selects the best solution in terms of function, material, cost and economy of manufacture. The design process can be illustrated by means of a chart as shown in Fig. 8.1.

The material selection and economy of manufacture will decide the design form and the lowest cost design. The best functional design will consider using the minimum number of components in order to meet the design objective and achieve the most effective operation in service. This should take account of the most economic design form depending on the methods of manufacture open to the designer. Figure 8.1 therefore shows a dotted line between lowest cost design and the best functional design boxes, indicating the influence of overall cost. Lowest cost always dictates the final design form. If the cost is above the budgeted amount, then it is necessary to re-examine the design to establish if a reduction in cost can be achieved by simplifying the component form. This may sometimes be achieved by eliminating unnecessary machining operations, increasing tolerances, reducing surface finish restrictions, modifying the form to make it easier to cast, forge, mould, etc.

8.3 Design Objective

The design objective can originate from many sources but the main ones tend to be formulated from a combination of foresight and market need. During a company's market analysis, investigation of one product may spring up an altogether unknown product requirement. This in turn may completely override the original plans for the first product. However, the accumulation of ideas can produce a highly marketable product.

Having established the product need, it is necessary to formulate a series of proposals. By this time the designer should have become deeply involved with the project and obtained a personal "feel" by which a list of possible solutions can be begun. The solutions must encompass not only the basic requirement of performing the design function, but must consider the other factors which may not be written into the original specification but which must be included to ensure its efficient operation. These factors may include

1. Size suitable for installation or handling.
2. Environmental or climatic influences.
3. Serviceability.
4. Ergonomic design.
5. Good appearance.
6. Noiseless operation.
7. Minimum weight.

8.4 Design Conception

At this stage the designer must not attempt to fine down concepts of the design, but keep an open mind so that ideas can flow more easily. At the initiation, creativity is at the highest level and the physical considerations are at the lowest. The design sketch should not be inhibited in any way by restrictions which the designer may think will impair its operation. In some companies the method of brainstorming is used where a group of people sit around a table and throw ideas into a melting pot where they can be used for accumulating ideas for use at a later stage. However, although it is thought that a group of individuals can produce more ideas this way, it is not necessarily possible to be original in design concept. Care must be taken to avoid negative attitudes towards the ideas put forward by this method, otherwise it tends to reduce the open-mindedness needed to make it a success.

8.5 Design Sketches

These should be confined to what some designers call A4 sketches. These are sketches of proposed ideas drawn freehand on pieces of A4 paper, which may be either plain or sectionally lined (i.e. graph paper). This is a most vital part of the design process, and each sketch should be carefully notated to indicate the important features. Figures 8.2–8.8 show typical sketches of a design for a mechanism which is required to meet the following design objective.

Objective A mechanism is required to actuate and engage a clutch. This may be carried out by means of a lever, or a handle which is operated by hand (see Fig. 8.2). Further, the shift fork must be

1 Attached to a shaft and be securely fixed.
2 Unable to move from its set positions A and B when actuated.
3 Able to move easily and require minimum effort to engage.

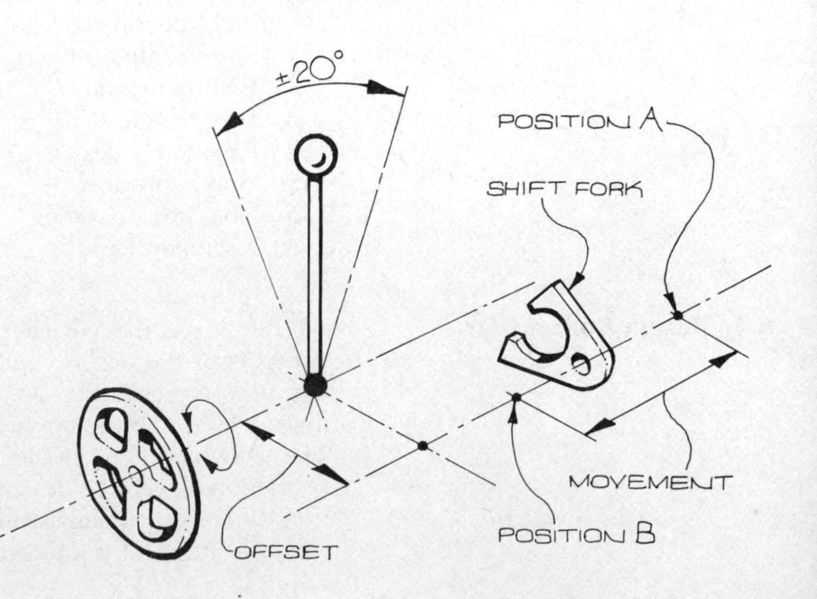

Fig. 8.2

8.6 Design Analysis

As stated earlier it is essential that some method is adopted to evaluate each design, and a systematic approach is recommended here. This may take the form of using a points system for each feature of the design specification. This may then be combined with another points system which takes account of method of manufacture.

By using a systematic design appraisal the optimum solution can more easily be found, so that

1 The minimum number of parts are used.
2 The minimum cost can be achieved.
3 The easiest method of assembly is adopted.

Table 8.1 shows how the points are allocated to the level of design need.

Having established the suitability of the proposed designs in terms of need, it is then necessary to relate them to suitability for meeting each design feature considered important by the designer, in order to meet the most economic and functional design. This can be done by allocating points as shown in Table 8.2.

Table 8.1

Level of design need	*Points (N)*
Design feature which must be included to ensure function.	4
Design feature which must be included to ensure service life.	3
Design feature which must be included but can be varied to suit design parameters.	2
Design feature which only needs to be included if physically and economically possible.	1
Design feature which is not essential to meet design need.	0

Table 8.2

Suitability	*Points (S)*
Optimum suitability	3
Nearly suitable	2
Marginal suitability	1
Unsuitable	0

Table 8.3

	8.3		8.4		8.5		8.6		8.7		8.8		*Ideal*	
Design feature	$N \times S$		$N \times S$		$N \times S$		$N \times S$		$N \times S$		$N \times S$		$N \times S$	
Positive function	4	3	4	2	4	2	4	1	4	2	4	1	4	3
Simple layout	2	3	2	2	2	2	2	1	2	1	2	1	4	3
Low inertia	3	2	3	2	3	2	3	1	3	2	3	1	4	3
Min. lost motion	3	3	3	0	3	2	3	0	3	0	3	0	4	3
Efficient action	3	3	3	1	3	2	3	1	3	1	3	1	4	3
Low friction	2	2	2	0	2	1	2	0	2	2	2	0	4	3
Ease of manufact.	2	2	2	0	2	2	2	0	2	2	2	0	4	3
Ease of assembly	2	2	2	0	2	3	2	0	2	2	2	1	4	3
Serviceability	2	2	2	1	2	2	2	1	2	3	2	1	4	3
Low cost	2	3	2	1	2	2	2	0	2	1	2	0	4	3
Totals	64		25		50		14		39		16		120	

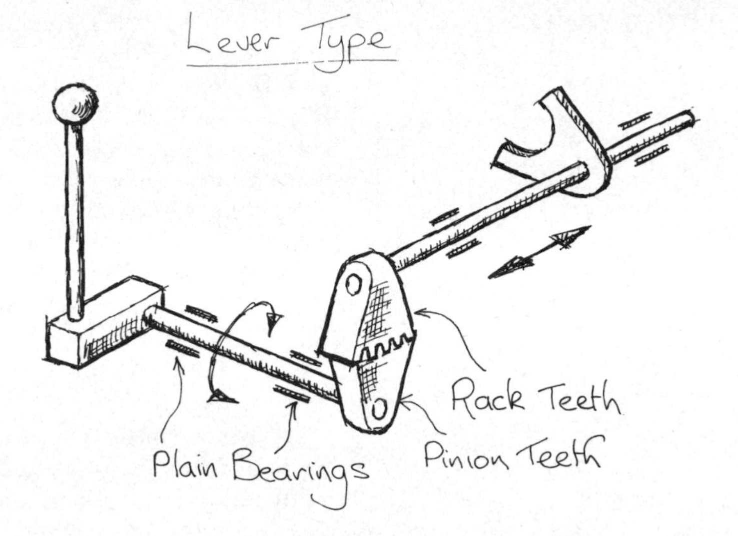

Fig. 8.3

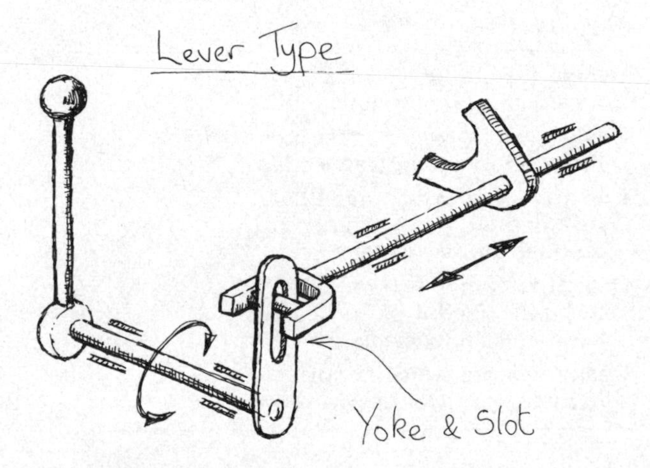

Fig. 8.4

The points from Tables 8.1 and 8.2 are multiplied together against each design requirement and added together to obtain the total points (Table 8.3).

Figures 8.3–8.8 show six alternative designs which would satisfy the design objective in Fig. 8.2. By adopting the systematic points method of design given above then it is possible to select the optimum design in terms of design, manufacture, assembly and cost. Table 8.3 shows how the total points are allocated to establish the optimum design.

The result of the analysis using the systematic points method gives the best solution. The designs of Fig. 8.5 and Fig. 8.7

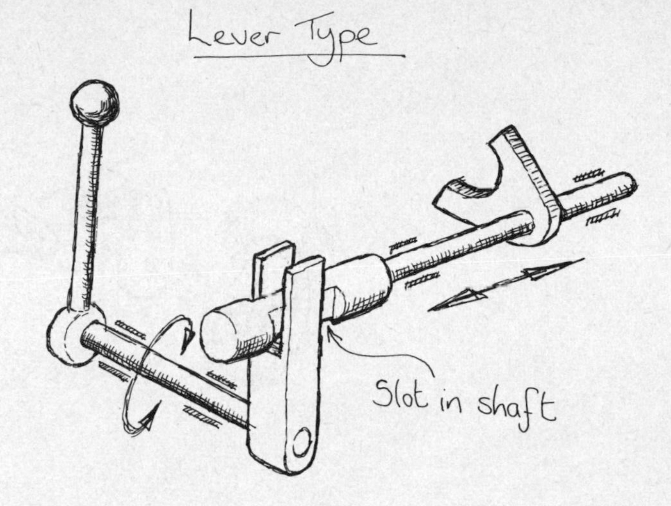

Fig. 8.5

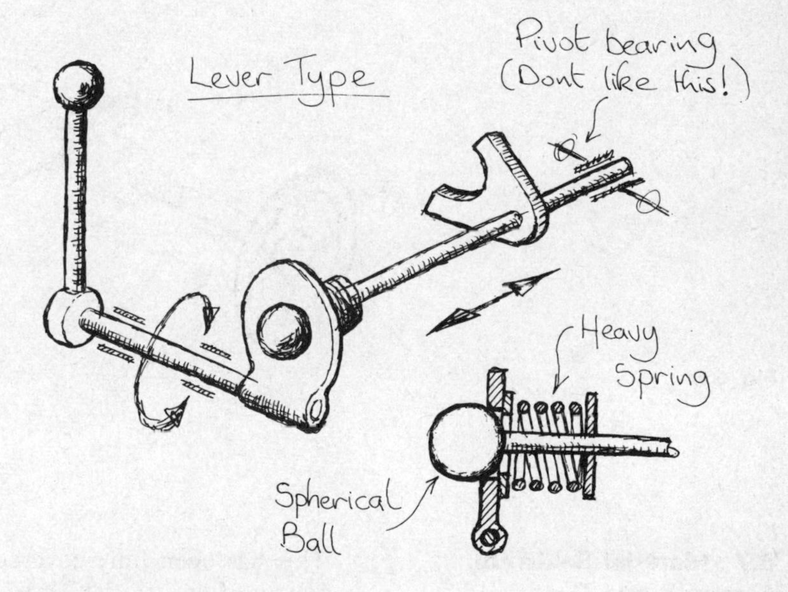

Fig. 8.6

would give reasonable solutions to the design objective, should the segment gear method be unsuitable from machine availability point of view.

The analysis should take into account the mounting of the parts within the housing containing the mechanism because the pins, bushes and drilled holes, etc., can raise the cost considerably.

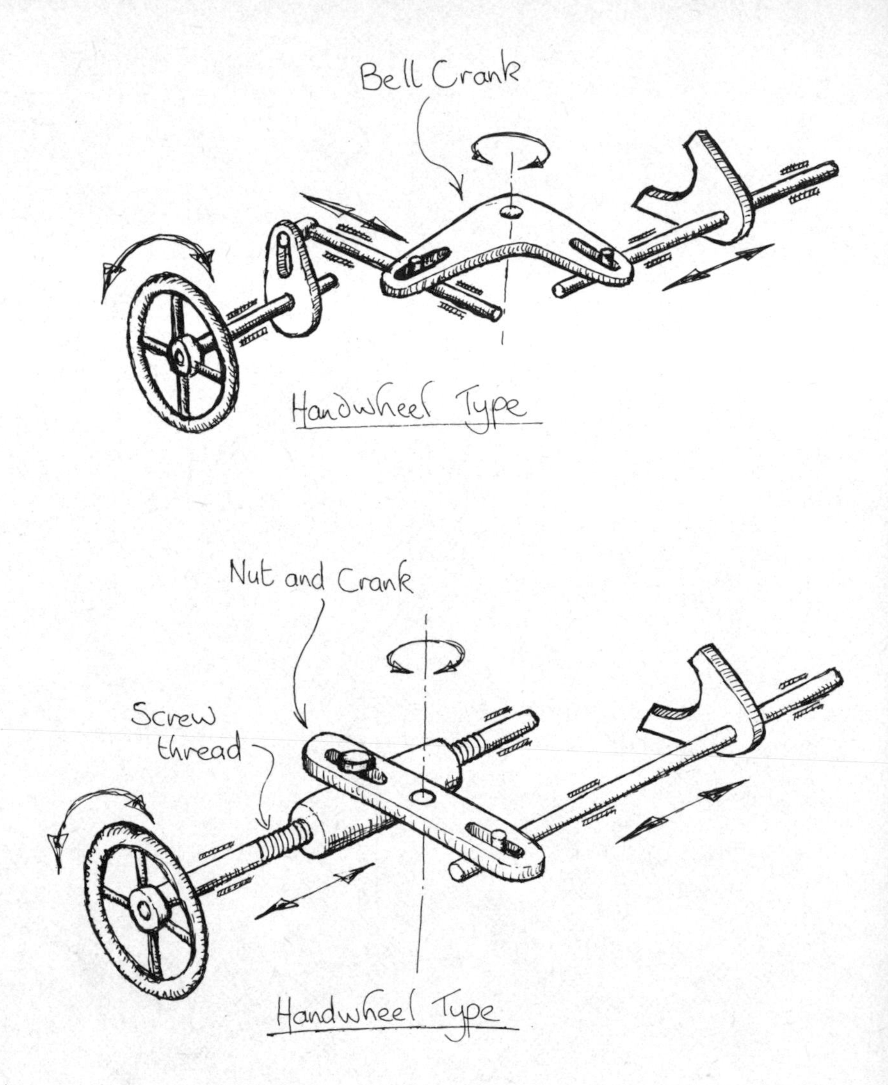

Fig. 8.7

Fig. 8.8

8.7 Material Selection, Strength and Cost

This has been fully covered in Chapters 2 and 3 and it only remains here to emphasise that the proposed design must be strong enough to support the envisaged loads. Strength is closely associated with efficiency and function, for if parts deflect or are inadequate in terms of stability then it cannot be considered a functional design and its operating efficiency will decrease because of high frictional loads.

Strength analysis should be carried out immediately the design has been selected. This involves establishing the physical properties of the materials (see Chapter 1) since thermal and chemical aspects of the application may have the effect of reducing the life of individual components. Ideally all the component parts should have the same life. However, this is not always possible due to the variation of friction which occurs between moving parts. This ultimately leads to wear and may cause seizure of the mechanism. Because of this it is essential that the parts are dimensioned to give reduced loading and easy movement between the contacting surfaces.

In order to keep the design to a low cost, the designer should tabulate the proposed materials and physical properties for each component part, and then establish the relative cost taking into account the density of the material. The method shown in Chapter 3 can be used for this.

8.8 Economy of Manufacture

Having established the design and subsequently the material to be used, it is necessary for the designer to decide on the method of manufacture to be used. This should initially be confined to the machinery available within the designer's own company.

The form (shape) of the individual component parts should be kept as simple as possible. This can be achieved by minimising the amount of machining and keeping tolerances to acceptable limits. Excessive use of geometric tolerances adds cost to a design due to the need to provide special fixtures and gauges. This in turn requires the need to have inspection procedures that add considerably to the cost of producing the part.

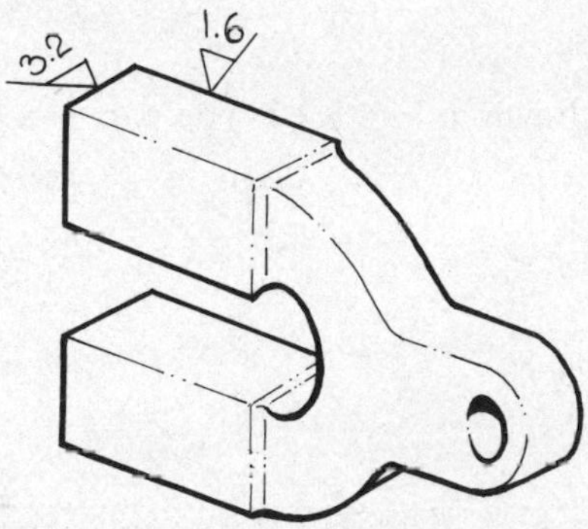

When designing castings or forgings that will only require a minimum of machining, it is advisable to raise those areas that will need to act as locating surfaces. Figure 8.9 shows typical illustrations of this as applied to the shift fork to move the clutch. Where the surface is in direct contact with another moving part, it is necessary to have a good surface finish. On other surfaces requiring machining, but that are not in contact, then the surface finish can be relaxed.

Fig. 8.9 Shift fork for mechanism

Parts of the mechanism which are slotted should be fine blanked if possible, as this will eliminate any subsequent machining operation. The quantity of parts to be produced will determine if this is economically feasible.

Where knobs or handles are to be incorporated in the design, then it is advisable to consult manufacturer's catalogues which specialise in these parts. This may also apply to other parts such as springs, dowel pins, handwheels and bearings.

The outcome of this design analysis should result in the most economic design in terms of function, material and manufacture.

8.9 Re-design Analysis

A designer is not always fortunate in being able to design a new product from a new basic specification. In many instances the designer's skills are confined to re-designing or modifying an existing company product. Although this can sometimes be a tedious task, it is not always an unrewarding one. It can, if the designer is ingenious, result in a product that has an increased consumer demand, which in itself increases the designer's awareness of his ability. This will have the effect of ensuring that the designer will always produce designs of a high quality, drawing on the experiences of past successes.

The marketing department of a company has the responsibility of ensuring that the product which it sells is competitive in the market place. To a great extent this depends on the designer's ability to choose the correct design features which will keep the cost of producing it to a minimum. When re-designing a product the designer must look at each component to see if its design can be improved in such a way that it will be less costly to produce.

This can be achieved in several ways:

1 Reduce the volume of material used in its construction.
2 Change the material to a less costly one.
3 Change the method of manufacture.
4 Change the design form.
5 Eliminate unnecessary close tolerances.

To illustrate how a component may be modified so that a cost saving is achieved, consider the following example.

Example

A casting for a fluid pump is shown in Fig. 8.10. The part is to be made in large quantities.

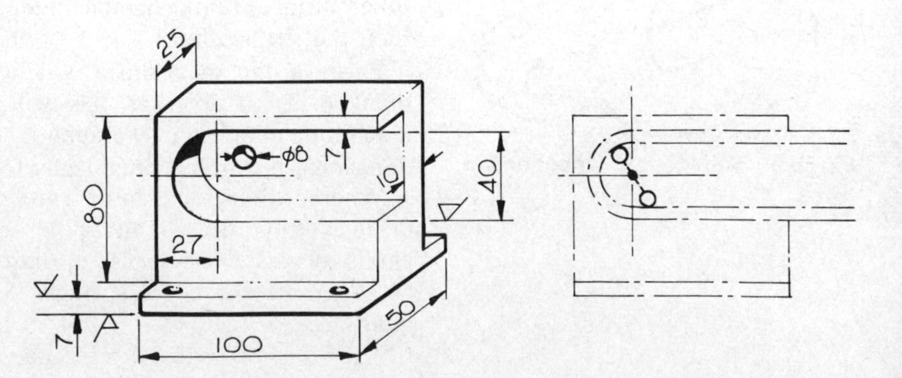

Fig. 8.10

The existing design consists of an iron casting which is machined to a close tolerance internally to ensure that the pumping action is uniform. The casting is mounted by means of feet which are machined both sides so that the part sits squarely on its mounting. The top faces of the feet are also machined so that the fixing bolts can sit squarely to their axes.

It will be apparent at first glance that the mass of the part is very high adjacent to the cavity containing the pumping mechanism. By reducing the volume of material around the casting, material cost can be saved. However, in doing this the stiffness of the part should not be impaired. By placing webs in strategic positions it is possible to maintain the casting's stiffness.

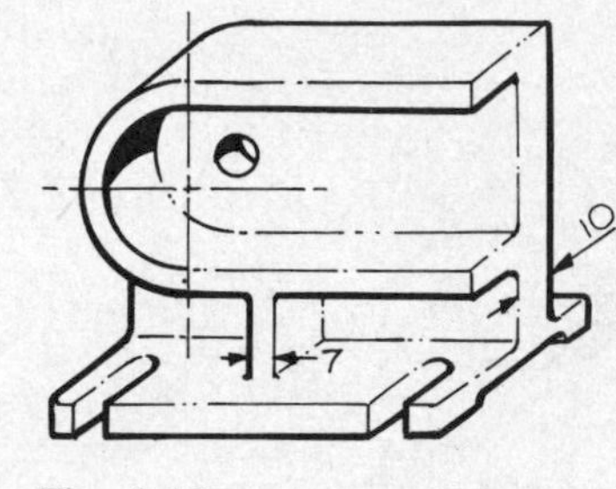

Fig. 8.11

Figure 8.11 shows one method of how this could be achieved using the same material. By calculation, the original volume of the pump body was found to be 182 cm^3. Taking the cost of grey cast iron as 20 p per kg, the material cost for the part is 25.7 p.

Comparing this now with the re-designed casting shown in Fig. 8.11, the volume is now reduced to 136 cm^3. The material cost reduces to 19.3 p which is a cost saving of 33%.

The new design would require a new pattern, but due to the fact that the part is to be made in large quantities the pattern cost would be recovered within a very short period.

By making the mounting holes slotted they can be "cast-in", thus saving the cost of drilling and special fixtures. Further machining costs can be saved by raising pads around the mounting holes underneath. This will increase the cutting tool life since there is less material to be removed.

If the quantities being produced are large enough then it may be less expensive to produce the part as an aluminium pressure die casting, which would result in the elimination of machining the mounting pads. Also, only a small amount of material would be required around the internal recess, thus keeping machining time and tool wear to a minimum. Although aluminium is approximately 3.4 times more expensive than grey cast iron, the cost of machining can be up to 3 times less than that of grey cast iron.

The above example has served to illustrate the means by which drastic cost savings can be achieved on an existing component. However, it should be emphasised that the designer should never produce a new design with the thought that, should it prove uneconomic to produce, then a re-design exercise can be carried out. The designer should, on the contrary, ensure that the most economic component design has been achieved at the outset of the original design.

8.10 Welding versus Casting

In applications such as housings, brackets, baseplates and complex covers, the designer is often faced with the choice of using a casting or a welded fabrication. The governing factor is, inevitably, cost. Because of the initial cost of producing the pattern, a casting will undoubtedly be more expensive on small production runs. As the required quantity increases, however, welded fabrications quickly become more expensive and make casting a favourable proposition.

Figure 8.12 shows how a component may be designed as either a casting or a welded fabrication. In both cases the component is subsequently machined and drilled.

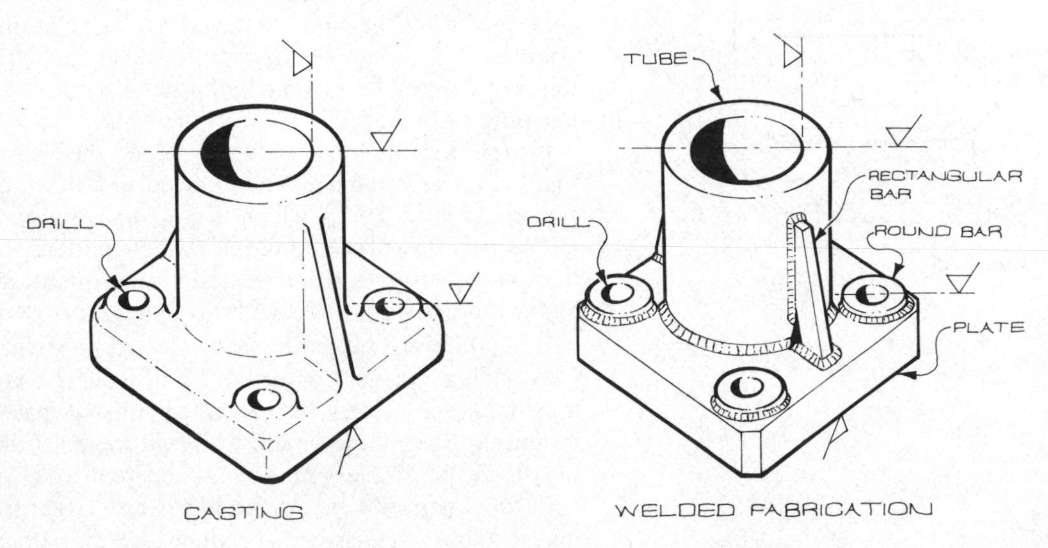

Fig. 8.12 Design as casting or welded fabrication

Problems

8.1 Figure 8.13 shows the design for a bearing pedestal. The existing design is manufactured from cast iron and has a solid form. It is required to re-design the bearing pedestal so as to produce a manufacturing cost saving. Using the same material, produce several design sketches and estimate the probable cost saving of each one. Take the cost of cast iron as 20 pence per kg. It should be remembered that the proposed designs should not impair the load carrying capacity of the part.

8.2 Figure 9.16 shows the design for a clamping mechanism used to hold a component whilst being machined. When installed it is found that the clamping action is not as efficient as it should be. Study the action required of the clamping mechanism and make suitable sketches of at least five alternative designs.

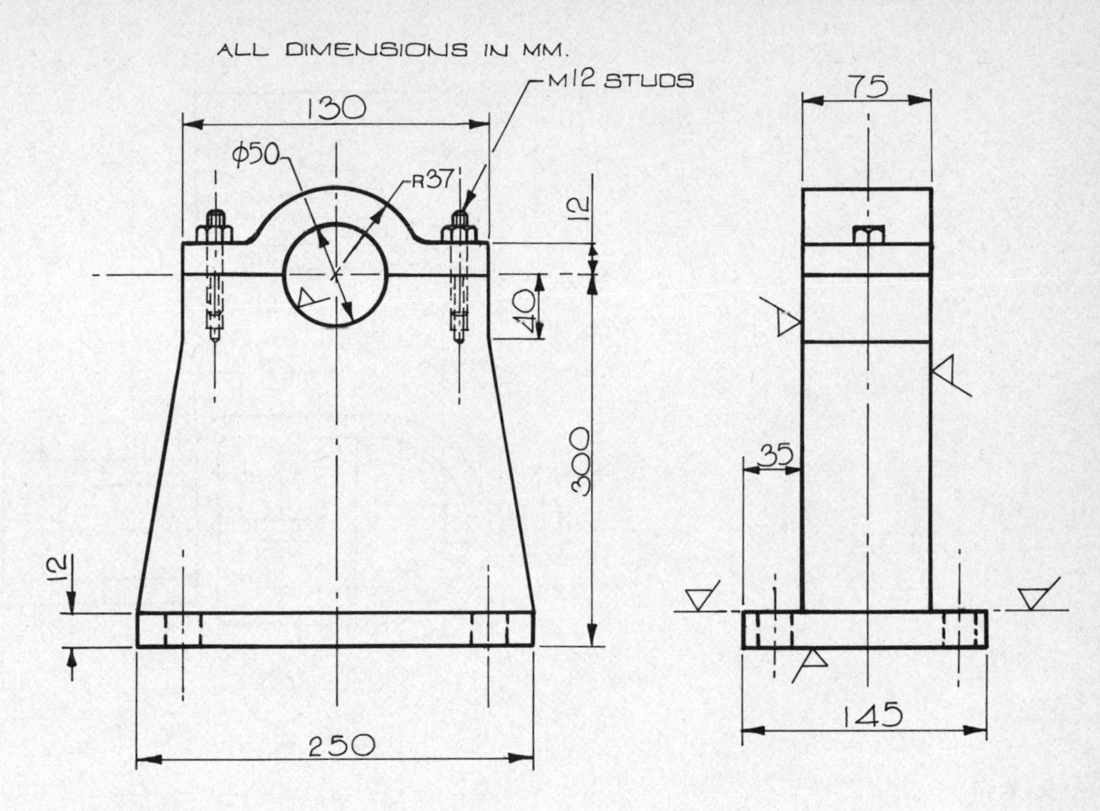

Fig. 8.13

Using Tables 8.1 and 8.2, carry out a systematic design exercise to find the most suitable alternative solution.

Prepare a chart similar to Table 8.3 and allocate points using the same list of requirements. i.e. positive function, simple layout, etc. Give a summary at the end to explain why you think the ideal solution obtained by this process is the best for the proposed duty.

8.3 A mechanical design is required to impart linear motion to a machine carriage. The following ideas were considered:

a) Lead screw and nut
b) Rack and pinion
c) Chain and sprocket
d) Hydraulic cylinder

Prepare sketches of suitable schemes based on each proposal and make a decision as to which is the most desirable solution by using a systematic procedure as described in the chapter.

8.4 Outline and briefly describe the important stages in the design process of a manufactured article.

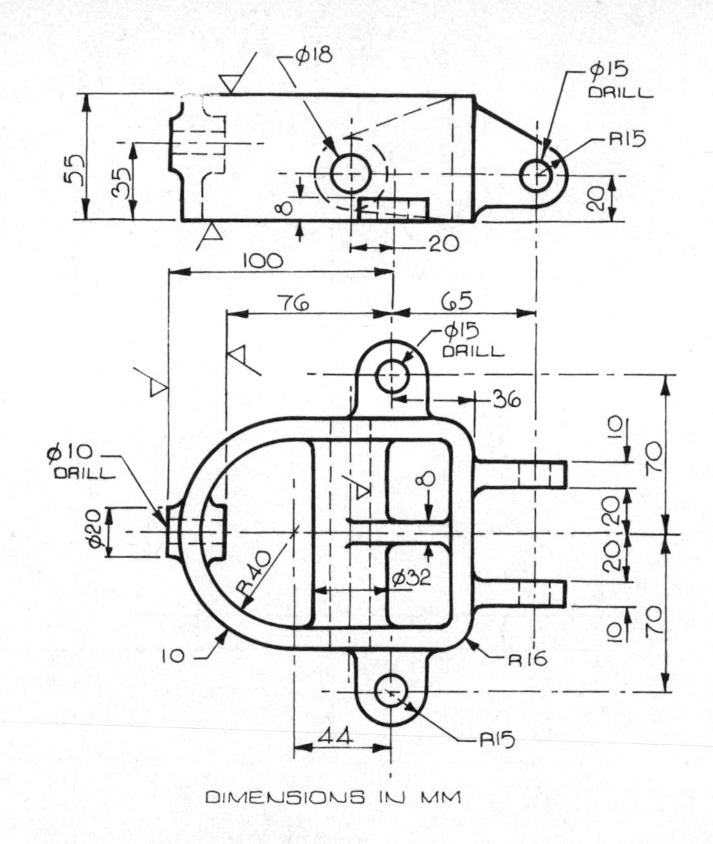

Fig. 8.14

8.5 Redesign the casting shown in Fig. 8.14 above as a welded fabrication of mild steel sections. Complete a detail drawing of your design, stating each type of weld used, with the appropriate British Standard symbols shown in Chart 1 (p. 138).

9 Mechanisms and Safety in Design

9.1 Mechanisms

A **mechanism** describes the way in which a machine works. All machines consist of at least one type of mechanism. With gears, the mechanism is the action of the gear teeth in contact. With a pulley drive, it is the frictional contact of the belt on the pulley wheel to drive it. With all machines the degree of efficiency depends on the type of mechanism employed, because the different types of element which make up the mechanism have different frictional behaviour when they function.

Mechanisms are used as a means of producing force or motion from one part of a machine to another. The mechanisms to be considered here will be screw threads, cams and linkages—gears having been covered in Chapter 5.

9.2 Screw Threads

Screw threads offer the designer one of the simplest and cheapest methods of converting rotational motion into straight-line motion.

All thread forms are based on the principle of a sliding tooth whose length lies along a helical curve around a cylinder. When considering the advantages of simplicity and the large speed ratios which can be achieved, the disadvantages must also be taken into account. These are, firstly, that transmission of motion is entirely rotational to linear and not vice versa, and secondly that motion is entirely a sliding action with the consequences of low efficiency and high wear rate. All transmission threads require adequate lubrication for satisfactory operation.

Some of the more common types of screw thread used in power transmission are shown in Fig. 9.1.

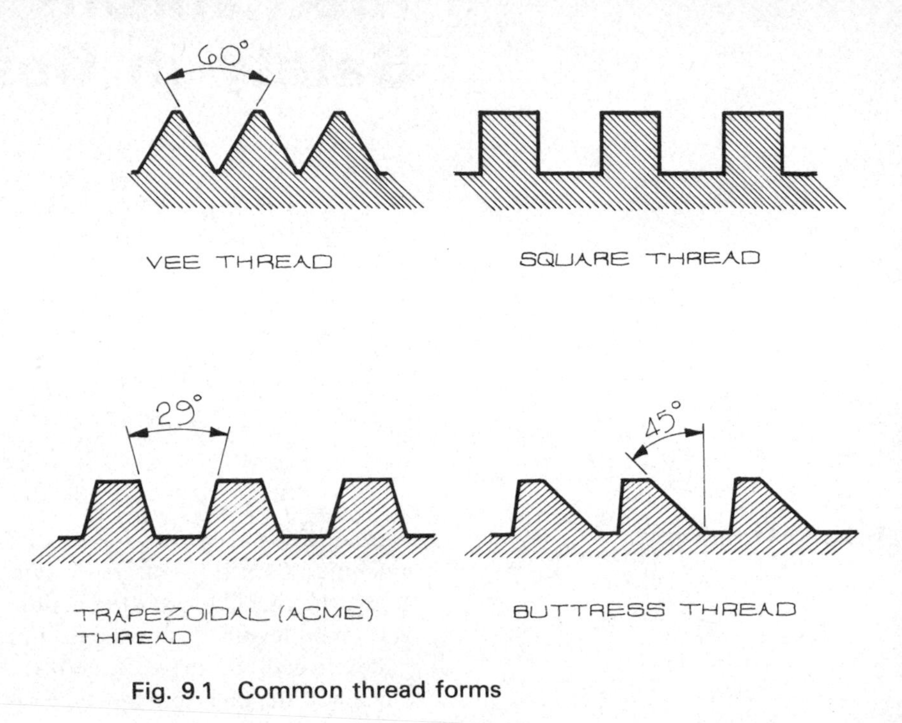

Fig. 9.1 Common thread forms

Vee thread Although chiefly used for mechanical fastenings, this type of thread is also used to transmit motion in lower-duty slow-speed applications.

Square thread This is used for heavier-duty transmission drives and will accommodate axial thrust loads. Square threads are commonly used in applications transmitting heavy vertical loads, such as screw jacks.

Trapezoidal (acme) thread This type of thread originated in the U.S.A. and is often used in feed screw applications. It will accommodate a higher thrust load than the square thread.

Buttress thread This will resist heavy axial loads in one direction only. The thrust-bearing face is nearly vertical whilst the back face slopes at 45°. Buttress threads are commonly used in quick-release mechanisms such as those in hand vices.

9.3 Cams

Cams are machine components that produce motion in a single plane, usually up and down. The cam revolves about a centre and the profile of the cam varies in distance from this centre. This produces a rise and fall of the follower during the cam's rotation, and the degree of this movement is designed to fit the special needs of the machine itself.

Cams can be designed to produce

1 Uniform or linear motion (uniform velocity)
2 Harmonic motion
3 Gravity motion (uniform acceleration)
4 Combinations of **1**, **2**, and **3**.

Some typical cams are shown in Fig. 9.2.

Displacement diagrams are used to analyse the movement of the follower relative to the rotation of the cam. In some instances it is necessary for the follower to remain stationary at some period during the rotation of the cam. This stationary period is known as *dwell*.

Fig. 9.2 Types of cam

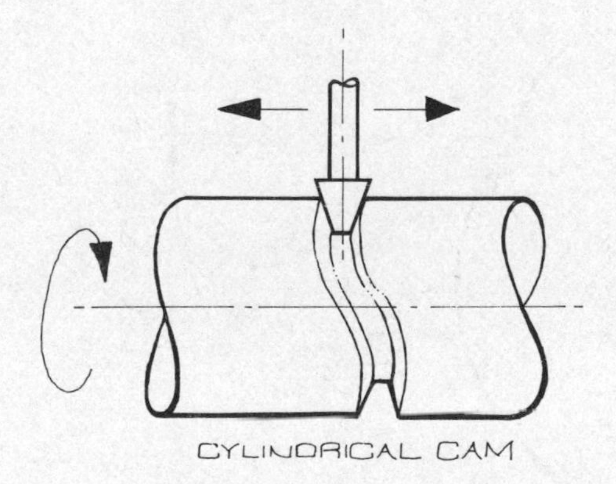

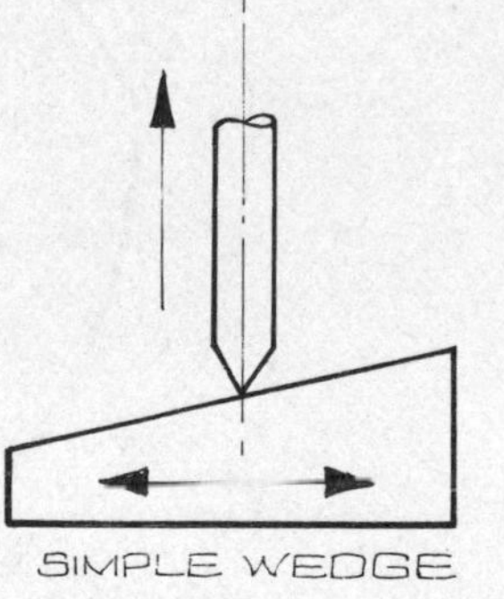

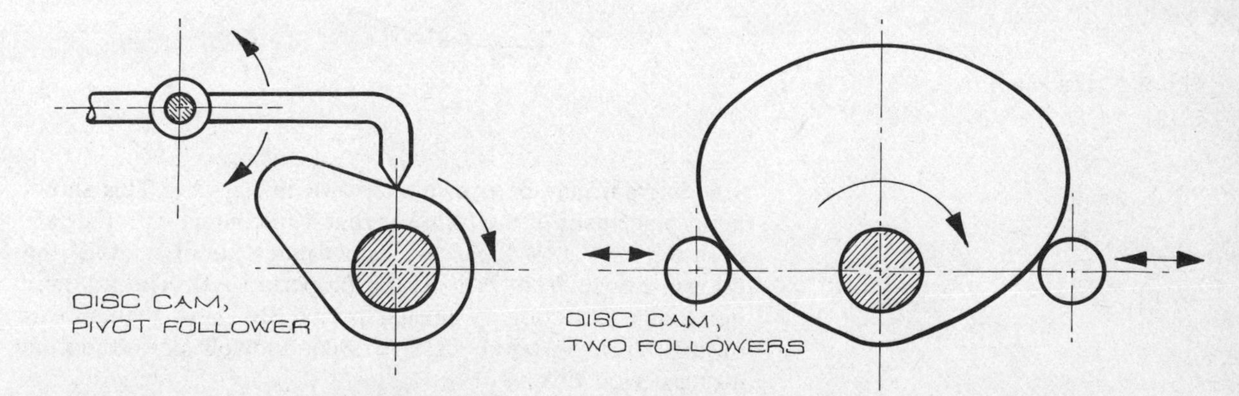

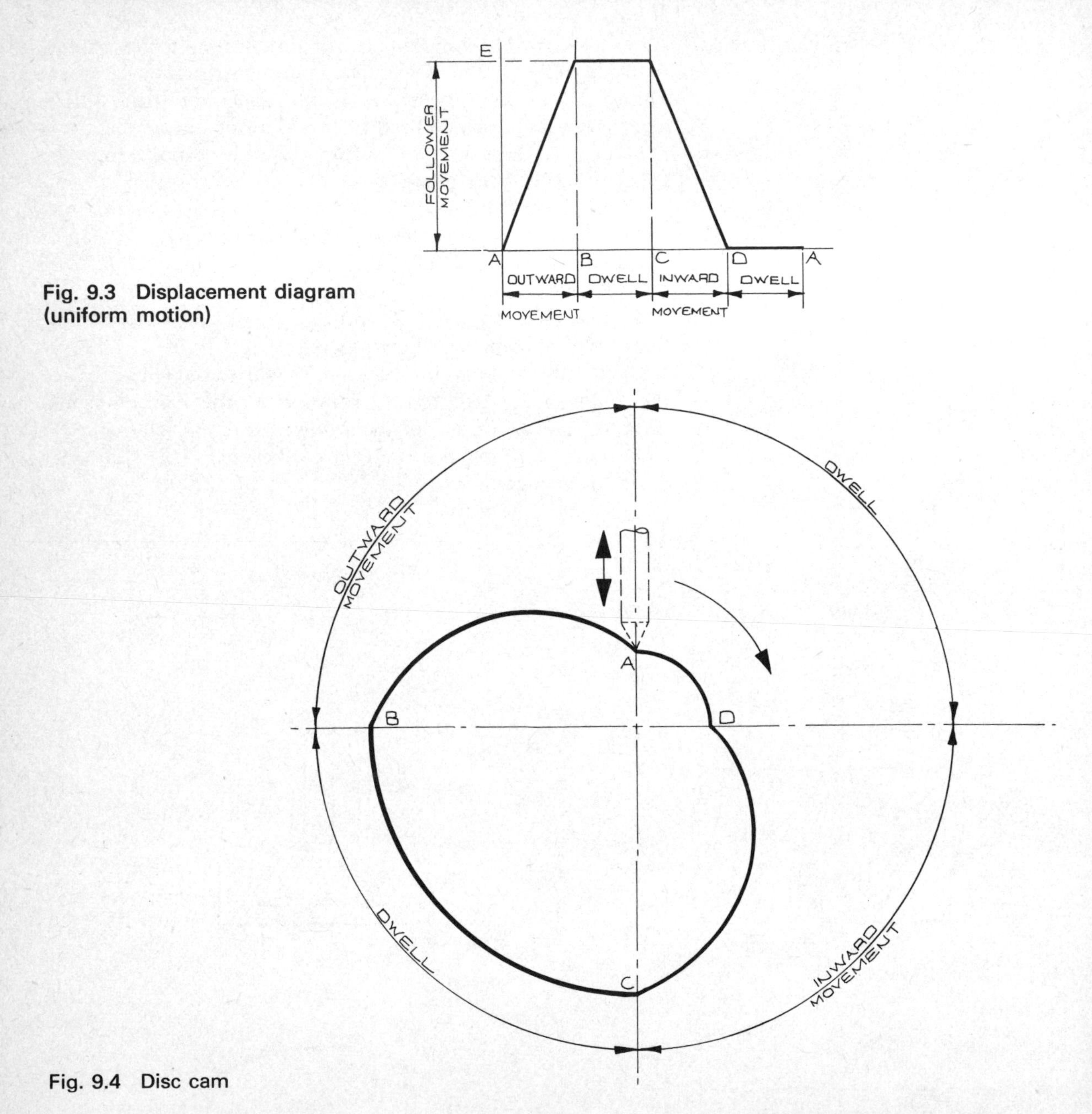

Fig. 9.3 Displacement diagram (uniform motion)

Fig. 9.4 Disc cam

A diagram may be drawn as shown in Fig. 9.3. This shows the displacement of the follower that is in contact with the cam as shown in Fig. 9.4. As the cam rotates about its axis, the follower moves from A to E during period AB. The follower then remains stationary during period BC, and then moves inwards during period CD. A second dwell period occurs during period DA.

9.4 Uniform Motion (Uniform Velocity)

The displacement diagram shown in Fig. 9.5 represents the follower movement when the cam rotates 180°. It can be seen that there are abrupt changes of velocity at two points. This is not practical since there is a tendency for the follower to move erratically and give uneven motion to the mechanism. This type of cam would only be suitable for very slow rotational speeds. It is normal to modify the cam profile so that arcs are placed at the points of change on the cam. The dotted line in Fig. 9.5 shows the modified displacement diagram when these arcs are introduced.

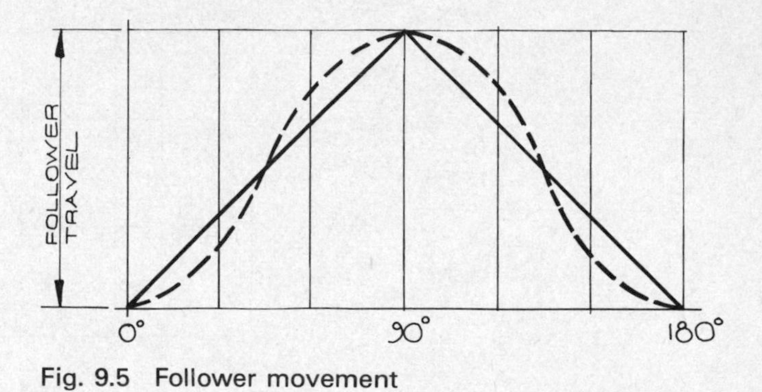

Fig. 9.5 Follower movement

9.5 Harmonic Motion

Figure 9.6 shows the displacement diagram for a cam giving simple harmonic motion to a mechanism. This action gives a smooth continuous motion when the cam is rotated at moderate speeds. The surface of the cam is made up of an even series of curves which results in the follower moving gradually from zero to maximum stroke in one revolution of the cam.

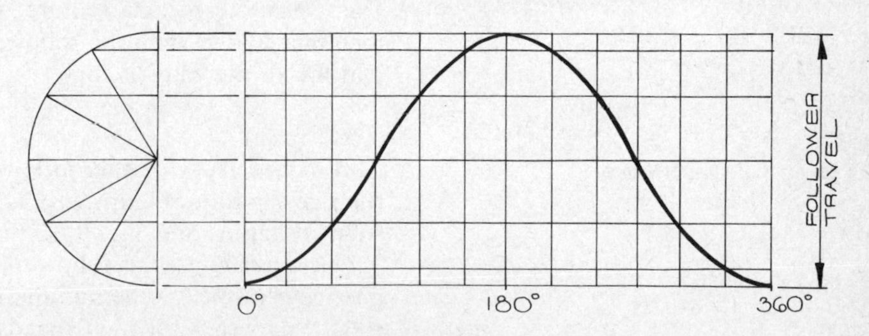

Fig. 9.6 Simple harmonic motion

9.6 Gravity Motion (Uniform Acceleration)

This type of cam profile is normally used for high-speed operation. The variation of displacement is analogous to the force of gravity exerted on a falling body, with the difference in displacement being 1; 3; 5; 3; 1 based on the square of the number. For instance, $1^2 = 1$, $2^2 = 4$, $3^2 = 9$. This same motion is repeated in reverse order for the remaining half of the movement of the follower.

The gravity fall of the follower is designed to conform to the shape of the cam so that its contact with the surface will provide smooth operation. Figure 9.7 shows the displacement diagram for a cam giving gravity motion.

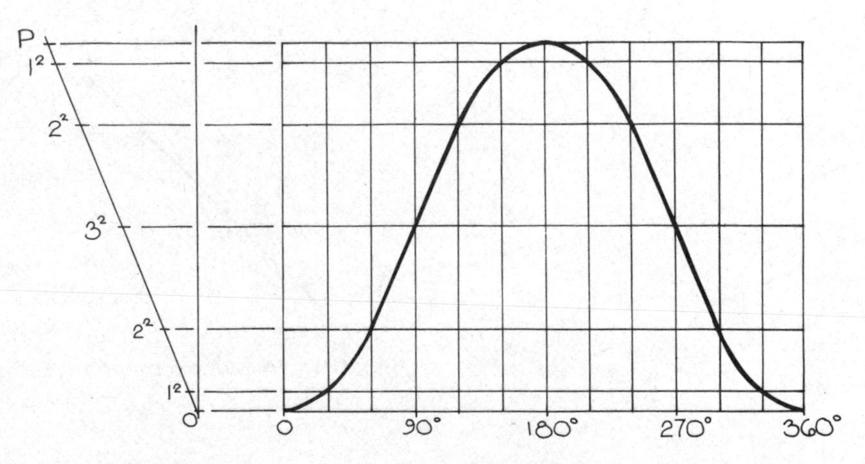

Fig. 9.7 Displacement diagram (gravity motion)

These types of cam are widely used in the internal combustion engine for operating valves. This is because the action obtained during the outward stroke gives a high acceleration to the follower, and on the inward stroke the acceleration of the follower is low, but the retardation is high. This ensures that the valves open and close quickly to give free flow of gases to and from the cylinders.

Because these cams require only a small external force to maintain contact with the follower, wear is kept to a minimum and life of the cams is high.

9.7 Cam Follower

Three basic types of **cam follower** are used. Figure 9.8 shows their design form. Figure 9.8*a* is the **flat** follower, Fig. 9.8*b* the **roller** follower, and Fig. 9.8*c* the **knife-edge** follower.

Flat and knife-edge followers are generally used for slow-moving cams where the minimum of force will be exerted. The roller follower is for high speeds and can transmit high forces to other moving parts.

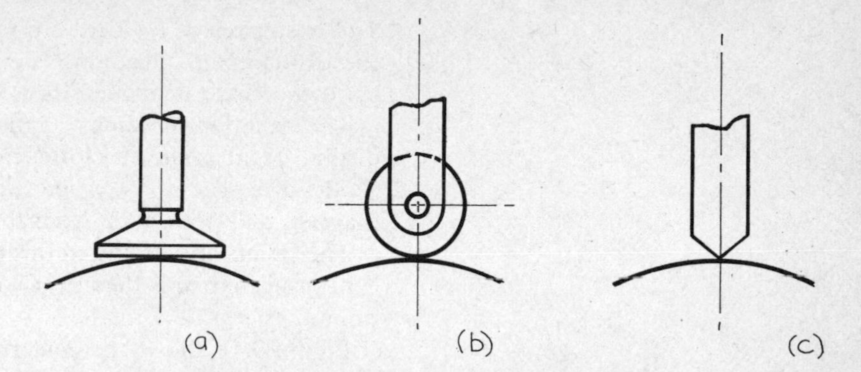

Fig. 9.8 The three basic types of cam follower: (a) flat; (b) roller; (c) knife-edge

9.8 Construction of Cam Form

The construction of a plate disc cam can be carried out simply by using the displacement diagram. Consider the example of a cam having harmonic motion and a knife-edge follower as shown in Fig. 9.9. The designer must know the following information before the cam can be constructed:

a) The desired motion of the follower, in this case harmonic.

b) The total rise of the follower.

c) The size of the follower and type, in this case knife-edge.

d) The position of the follower.

e) The base circle diameter.

f) The direction of rotation.

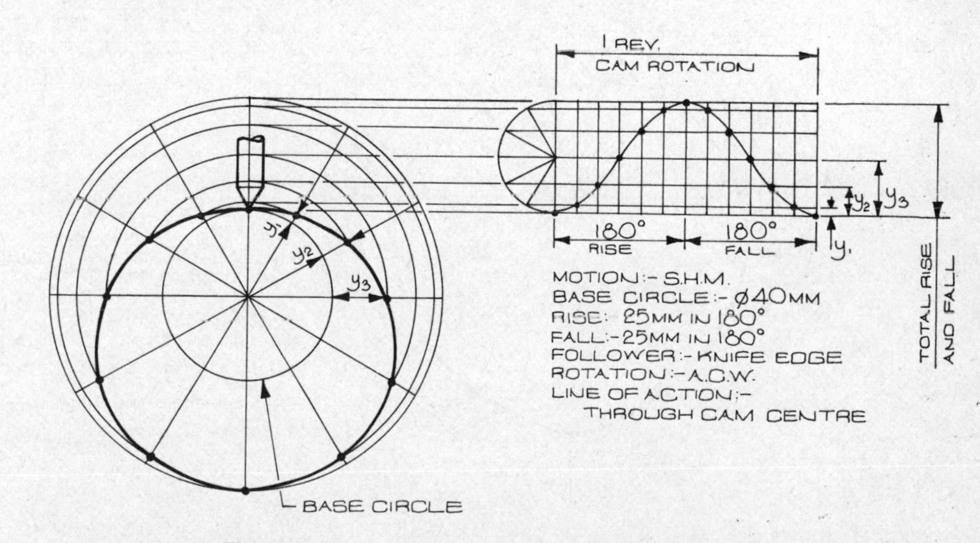

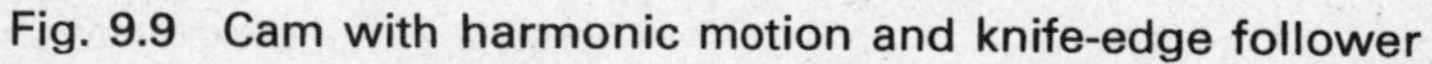
Fig. 9.9 Cam with harmonic motion and knife-edge follower

The base circle is divided into the same number of sections as the displacement diagram. There are twelve in this example, since the circle is divided into 30° sectors.

The rotation is shown anti-clockwise and therefore the displacement is shown plotted to the right of the follower.

The distances y_1, y_2, y_3, etc. are taken from the displacement diagram and positioned from the plate cam base circle.

The points are located in this manner all the way round. The cam profile curve is then neatly constructed to join the plotted points.

Figure 9.10 shows the construction of a plate cam to give a uniform velocity rise and fall to a roller follower. Usually roller follower cams require more accuracy and thus the circle and displacement diagram are divided into twenty-four (each division representing 15° rotation). Distances y_1, y_2, y_3, etc. are again measured from the displacement diagram and positioned from the base circle, but these distances must in this case include the roller follower radius, which is added to the amount of rise (i.e. y distance = roller radius + amount of rise). At each plotting, a circle is drawn of the same diameter as the roller follower. The cam profile is then constructed tangentially to the circles drawn.

Figure 9.11 shows a complex plate cam which will impart three types of rise or fall motion and two dwells for a knife-

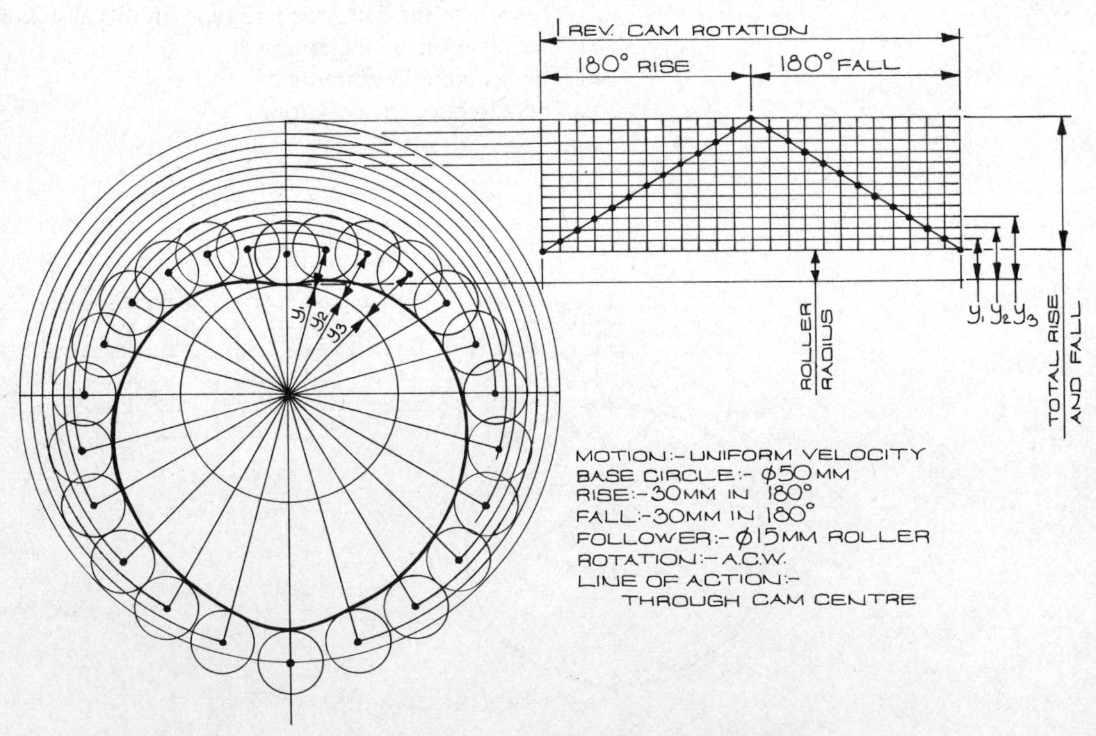

Fig. 9.10 Plate cam with roller follower

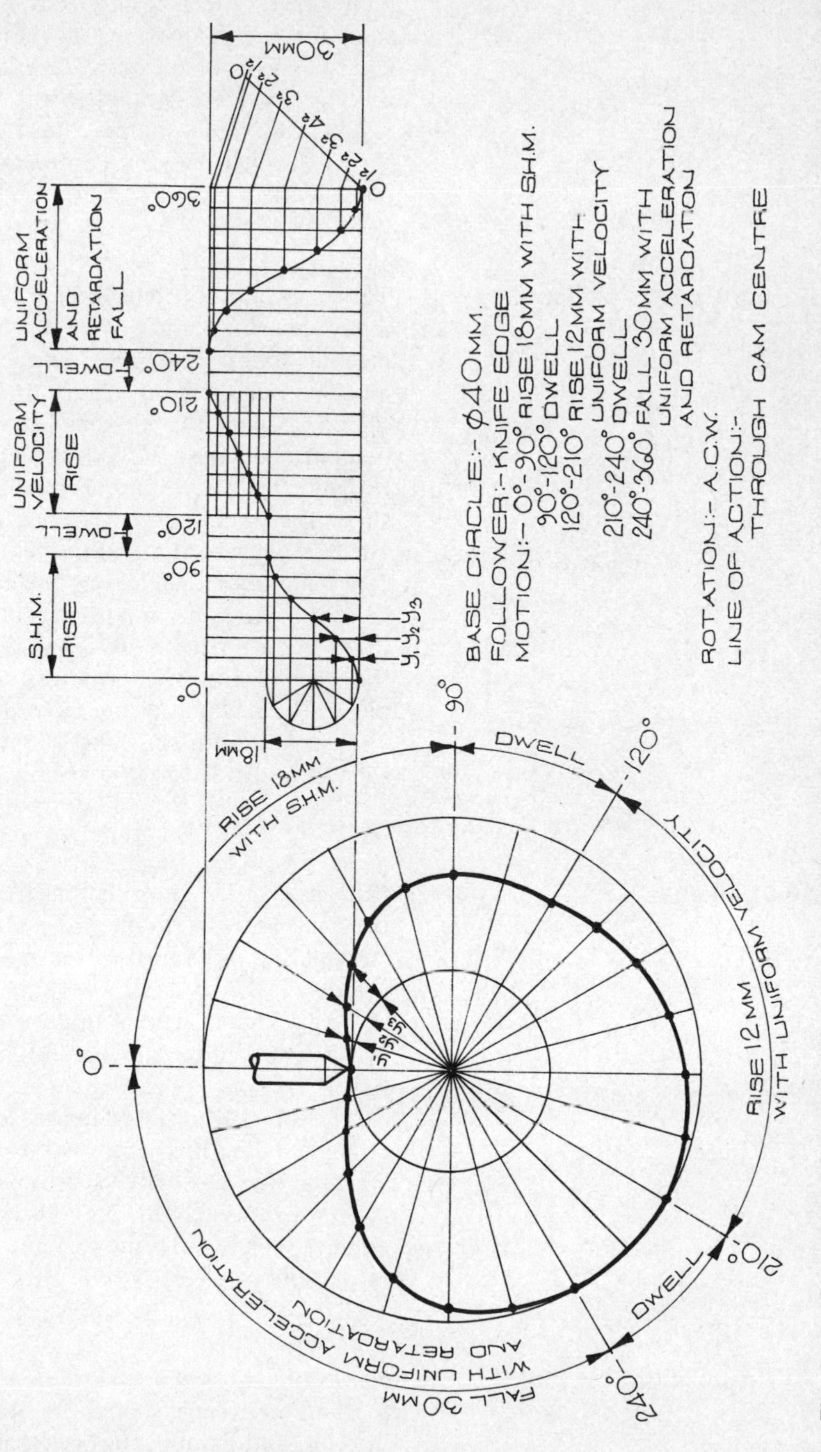

Fig. 9.11 Complex plate cam with knife-edge follower

edge follower. Here it is usual to divide circle and displacement diagram into twenty-four. Each type of motion has the same number of divisions on the circle as on the displacement diagram. For plate cams requiring clockwise rotation, the displacements are plotted to the left of the follower.

All plate cam examples here are considered for a line of action through the cam centre, although off-set followers are often used.

9.9 Application of Cams

The application of cams or cam type mechanism can be seen in many industries but one of the most common uses is in the machine tool industry. Automatic lathes used for turning small machine parts employ cams to index and feed tools. Other operations such as threading, undercutting and chamfering can be controlled with the aid of cams.

Figure 9.12 shows part of a machine tool for moving a turret or slide. The motion depicted shows the lead cam (1) rotating in an anticlockwise direction. This in turn has a roller (2) (the cam follower) moving along the cam profile, and as it rises and falls, it rotates the quadrant arm which moves the gear teeth (5). These gear teeth are in contact with a rack (4) attached to the machine turret, and the desired motion of the turret takes place. The reaction of the quadrant arm is taken by the pin (3) attached in the mainframe of the machine. The return of the slide is controlled by the spring (6).

9.10 Linkages

Linkages consist of struts that transform motion from one part of a machine to another. This can, for example, consist of straight line motion to rotary motion and back to straight line motion.

A very simple type of link mechanism is shown in Fig. 9.13. This shows that the movement of the lever transmits motion to link B.

One of the most common applications of a link is the straight sliding link as shown in Fig. 9.14. This is the form in which a slide is usually used to replace a link. This type of link is used extensively in the construction of internal combustion engines. Figure 9.15 shows some other links that can be found in machinery.

9.11 Application of Links

Many linkage systems are used in the design of clamping mechanisms. Typical of these are those that are incorporated into jigs and fixtures and used for clamping components being machined. An example of this device can be seen in Fig. 9.16.

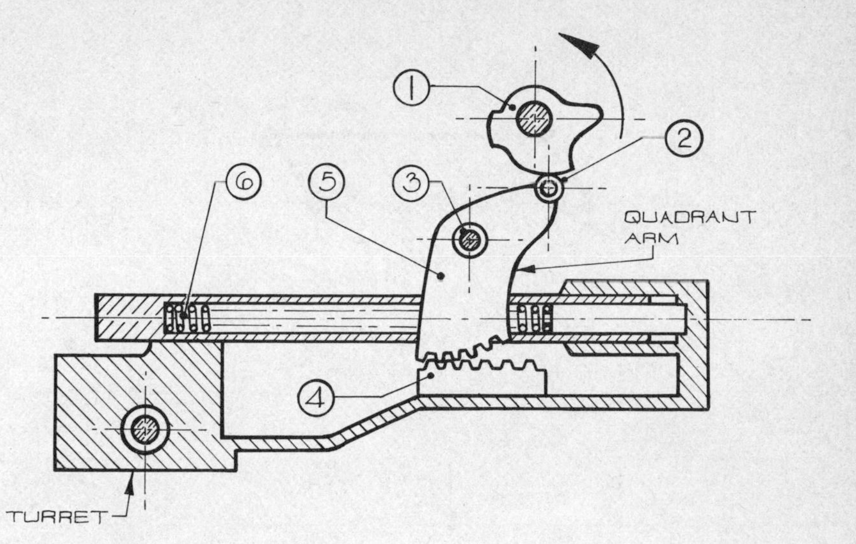

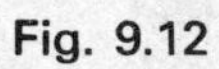
Fig. 9.12

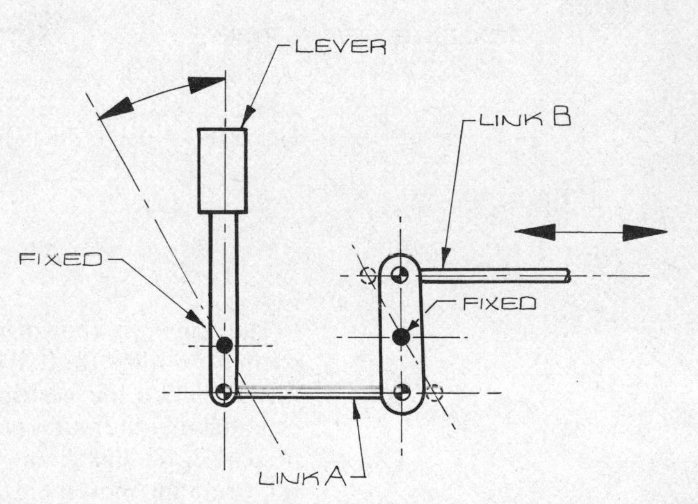

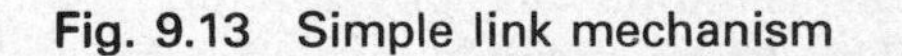
Fig. 9.13 Simple link mechanism

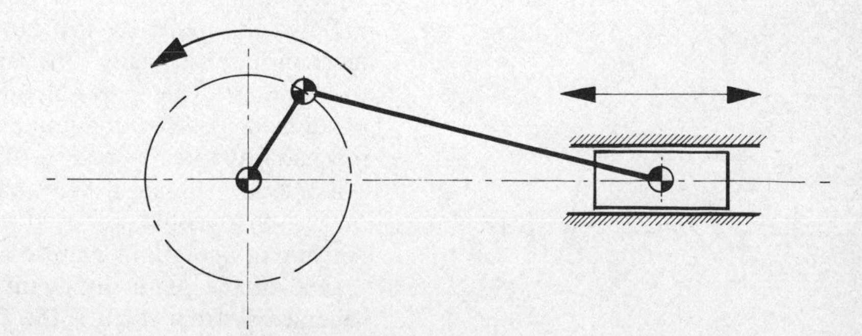

Fig. 9.14 Straight sliding link

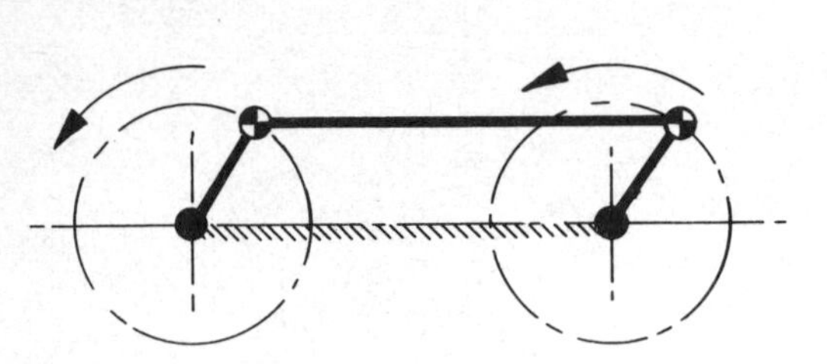

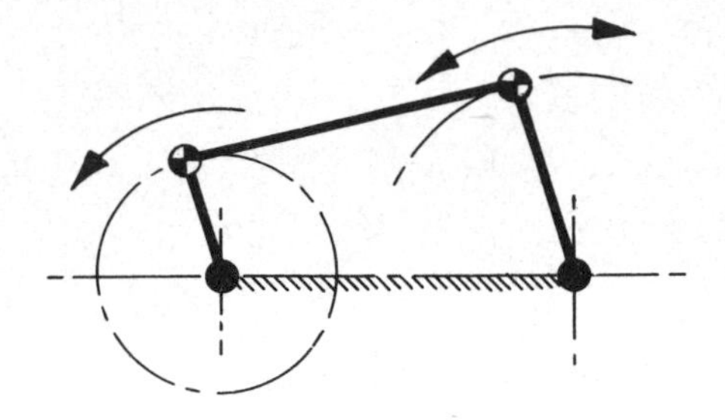

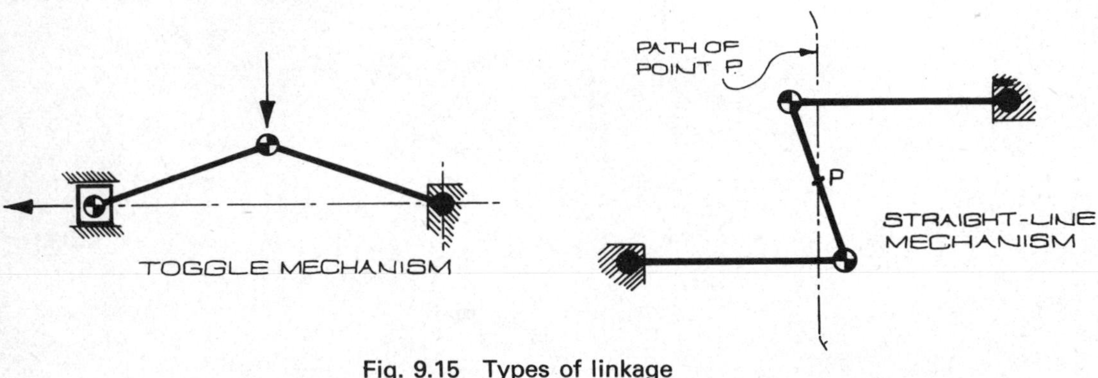

Fig. 9.15 Types of linkage

The handle is shown in its clamping position A and in its rotated position B. It is possible to rotate the handle even further until it touches the surface table it is mounted on. It is essential that the worst position is always drawn to ensure that no part of the linkage mechanism fouls, and therefore permits full operating movement. The designer should take account of hand positions and the available clearances. The design of the handle for operator comfort is a primary consideration for this type of equipment.

Linkages are used for all types of mechanical handling mechanisms, especially for transferring parts from one position to another. The part having been transferred could be positioned, fed or manipulated by other linkage-operated mechanisms to carry out further operations on the piece part. Automatic clamping devices based on the principles illustrated in Fig. 9.16 could be used then to hold the work so that machining operations can be carried out.

One of the main disadvantages of using mechanisms with several links in a chain is the slackness that is necessary at the

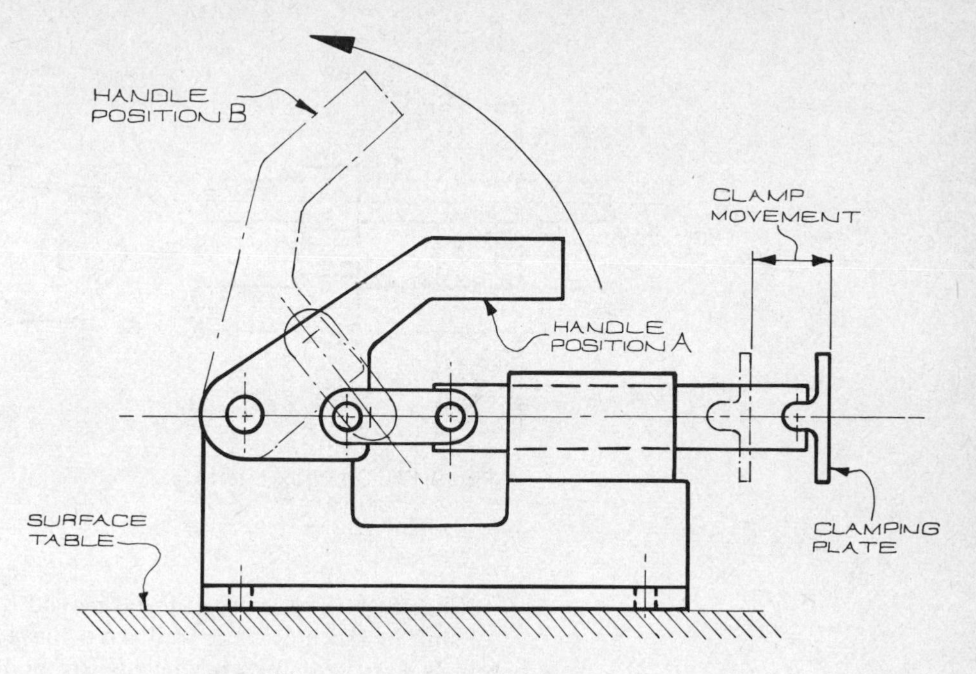

Fig. 9.16 Clamping mechanism

pivots to ensure free movement of the link. This slackness can be multiplied considerably from one end of the mechanism to the other, depending on the ratio of link arm lengths. The amount of lost motion resulting can be quite significant and a 0.1 mm slackness at one end could be 10 mm at the other. If universal joints are used in the mechanism as well, the resulting movement can be prohibitive. It is therefore essential for the designer to check this feature before finalising the design scheme.

9.12 Interlock Arrangement

There are many instances in engineering where it is necessary to ensure that one part of a machine cannot operate when another sequence is being carried out. In this instance it is necessary to introduce a device called an **interlock**. For example, in a gearbox for a motor vehicle it is essential to prevent the selection of two gears at the same time, otherwise severe damage can occur to the gears, shafts, etc. Figure 9.17 shows a typical solution to this problem.

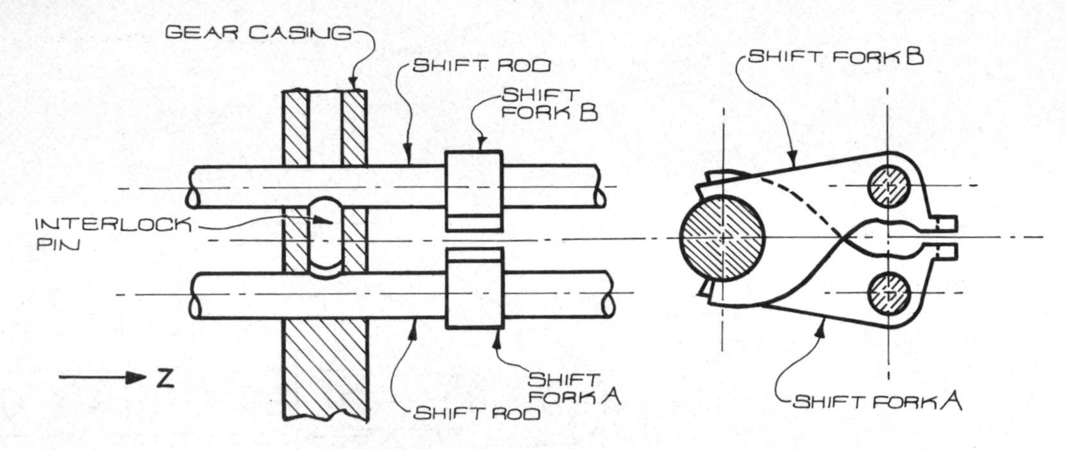

Fig. 9.17 Gearbox interlock

Shift fork A is required to be moved without the possibility of shift fork B moving as well. To achieve this a small interlock pin is designed into the mechanism and fitted into the gear casing as shown. Two slots are machined into the side of each shift rod which conform in shape to a radius on the end of the interlock pin. As the shift rod A moves in direction of arrow Z, the interlock pin is forced out of the slot in shift rod A and into the slot in shift rod B. This then locks shift rod B, through the interlock pin, to the gear casing and prevents its movement. Before shift rod B can be moved, the slot in shift rod A must be repositioned in line with the interlock pin. Selection of the gear shift rod B can then be made and the interlock pin is then forced into the slot in shift rod B which locks this rod to the gear casing.

Interlocks can also be used in electrical equipment to prevent connection of power to other sources when a particular circuit is in operation. The principle adopted is the same as that of the gearbox interlock.

9.13 Safety Devices

Safety devices are used for protecting machinery against damage due to

1 Faulty mechanisms
2 Incorrect operation
3 Mechanical overloading
4 Machine seizure.

The designer of machinery, whether rotating or electrical types, must assess the need to provide adequate safeguards to prevent damage should one of the above conditions occur. These safeguards are either mandatory or introduced due to the designer's own experience of the type of fault which could occur. Some of the different types of safety device which are commonly used are given in Table 9.1. These are split into two groups, the mechanical type and the thermal/electrical types.

Table 9.1

MECHANICAL	THERMAL/ELECTRICAL
Shear pins.	Fuses.
Slipping clutches.	Fusable plugs.
Shear grooves.	Thermostats.
Rotating governor.	Magnets.
Relief valves.	Visual & audio warning.
Timed delay.	Limit switches.
Tapered wedge.	Photo-electric cell.

The cheapest types of safety device are **shear pins**, **shear grooves** and the **tapered wedge**. The shear groove and shear pin are typically used in small pump drives which may be directly connected to a large machine. For example, the fuel injection pump of a diesel engine is driven by gears from the engine crankshaft. If the pump should seize for any reason, considerable damage could occur to the internal engine parts. To overcome this problem the pump drive shaft, or pumping rotor, has a groove machined in that is designed to torsionally shear at a predetermined load.

Spring-loaded hydraulic relief valves may be used in machine control systems so that, should pressure of the oil increase beyond the designed limit, then the oil is directed back to the tank and the system pressure stabilises to a safe limit.

Slipping clutches are popular in machine drives where instant automatic resetting of the drive is necessary once the overload has passed, for example in large gear drives driving conveyors. If it is anticipated that the overload will continue for long period, then the slipping clutch has a mechanical disconnect mechanism which must be manually reset before the machine can be restarted. This method prevents overheating of two slipping surfaces in contact.

In the case of thermal safety devices the most common one used is the **thermostat**. This is used in such equipment as heat treatment furnaces, plating baths, hot water tanks, etc. They are relatively easy and cheap to install and temperatures can be controlled between very close limits if necessary.

Fusable plugs are made from a wax that is designed to melt at a predetermined temperature. These devices are sometimes called waxstats. A typical example is the plug in a fluid coupling drive. Should the oil within the coupling become overheated due to high driving loads, then the plug melts and the oil is discharged, thus disconnecting the impeller and turbine. Other forms of waxstat are used in control linkage that are adjacent to heat sources. When the temperature rises beyond the design limit, the wax expands and operates the linkage.

Visual and audio warnings are commonly used in motor vehicles for letting the driver know if there is a malfunction of one of the vehicle's parts. For example, oil warning lights indicate a low oil level in the engine and the horn warns pedestrians or other vehicles of danger.

Limit switches may be used in machine tools for stopping a process at a predetermined point. For example, the table of a grinding machine strikes a limit switch which reverses the table motion and moves it back in the opposite direction.

Photo-electric cells are used as safety switching devices in applications such as press-brake guard interlock circuits, door-opening devices and burglar alarms. Referring to Fig. 9.18, light is allowed to enter the cell via a clear "window" and strike a cathode coating of light-sensitive material such as caesium or rubidium. This material then emits electrons to the anode collector and thus completes an electric circuit. If light is barred from entering the cell, the circuit is then cut off and the switch is brought into operation.

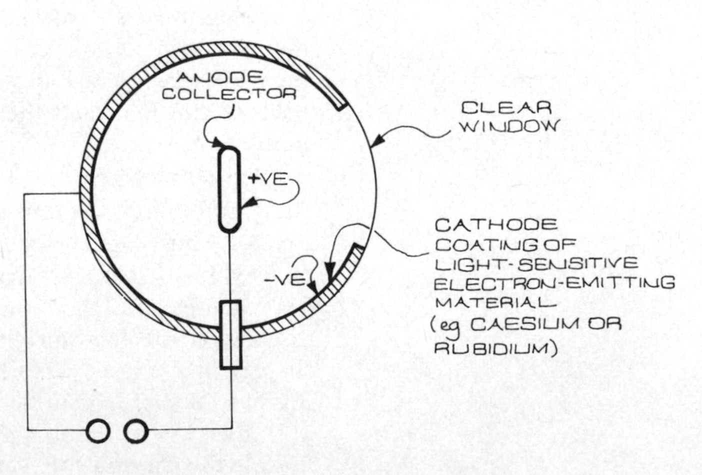

Fig. 9.18 Photo-electric cell

9.14 Machine Guarding

The designer of machinery has a social responsibility to ensure that a design is adequately guarded. However, in order to protect the consumer, requirements under part 2 Safety (General Provisions), section 12–16 of the 1961 Factories Act, states that machinery must be adequately guarded. The Health and Safety at Work Act, section 6 also states this.

BS 5304 Code of Practice sets out the principles of safely guarding machinery. This gives, under various headings, the types of guarding that may be used. Briefly some of these methods of guarding are

Interlocking guard One which prevents operation of the machine until a movable part uncovers the machine controls.
Automatic guard One which depends on the operation of the machine to operate it. The sequence of machining does not start until the guard is in position. No part of a person is exposed to danger.
Self-adjusting guard One which stops accidental access to a machine bay, but allows access of the workpiece.
Trip device One which indicates the approach of an operator and stops the machinery.
Overrun device A device used in conjunction with a guard that is designed to stop the machine if its exceeds designed speeds.
Fail safe design This is a method of safeguarding machine and personnel such that, should a malfunction occur, then it will fail to safety and result in the immediate stopping of the machine. Trip devices or overload clutches are typical of the solutions to this problem.

BS 5304 gives several other possible guarding devices and it is recommended that the designer is fully acquainted with these.

9.15 Common Hazards of Machines

All machines are a potential danger to the operator, and it can result in a person being

1. trapped bodily or having clothing caught up in the mechanism. Trailing belts, ties, sleeves, etc. may be the cause of this type of accident.
2. struck by a moving table or by a component coming loose.
3. cut by the tool being used to machine the component.

Rotating shafts, whatever the speed, are potential source of danger, as are projections connected to them. Fan blades, wheel spokes and gears have dangerous edges and guarding of these parts is essential.

Reciprocating machinery should never be placed in confined spaces, and it is recommended that guard rails are placed around the danger area.

Chains and belts should be checked periodically, even if guarded, to make sure that excessive wear has not taken place. These are able to burst through guards if the speeds of rotation are high.

Materials that are processed by moving at high speed must be firmly attached to the machine. Parts that are ejected automatically after machining must be covered throughout the ejection period to prevent them flying off the machine.

9.16 General Hazards

Other sources of danger that are not of a mechanical nature include

1. Hazardous materials.
2. Dielectric heating processes.
3. Noise and vibration.
4. Pressure and vacuum environments.
5. Ultra-violet (electric arc welding, ultra-violet printing processes)
6. Temperatures above normal.
7. Dust and grit.

The guarding of machinery and processes is an essential part of the designer's brief, and it is the designer's responsibility to take all reasonable care to ensure that the design is safe in operation.

The Health and Safety at Work Act 1974 makes all personnel responsible for anticipating risks to anyone, whether in the Company's employment or not. The Act makes employers criminally indictable if the Company has not taken all possible safeguards to ensure personal safety.

It should be noted that codes of practice by the British Standards Institution or other professional bodies are not approved under the provisions of the Health and Safety at Work Act. The commission set up by the Act is itself responsible for making arrangements for the implementation of the Act, and it can make use of the Codes of Practice to help it to do so.

Problems

9.1 Figure 9.19 shows the design of a clamping fixture. Prepare design sketches showing alternative methods of actuating the clamping plate. Consider using linkage, cams or gear mechanisms.

From these sketches carry out a design analysis using the systematic process, and establish the optimum design in terms of design manufacture, assembly and cost.

9.2 Design an interlock mechanism between the two discs shown in Fig. 9.20 such that when disc A is rotating disc B is stationary, and when disc B is rotating disc A is stationary.

In place of the handles shown to rotate the discs, re-design the assembly such that the discs are rotating by one gear that can slide between them. Redesign the interlock mechanism if necessary to achieve this new arrangement.

Draw the assembly full size and detail all relevant parts with sufficient information to manufacture it.

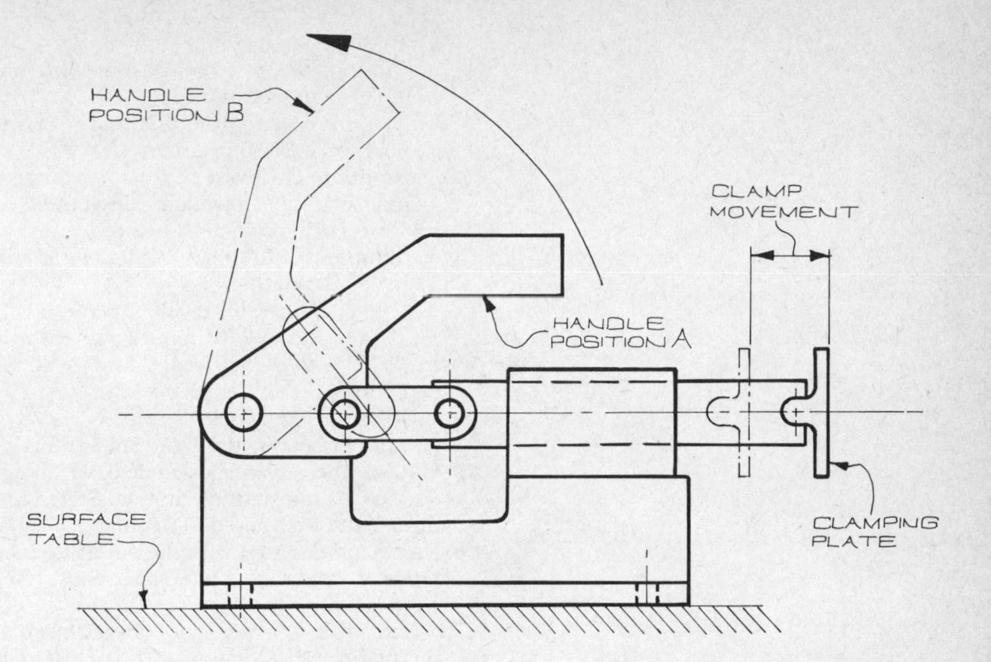

Fig. 9.19

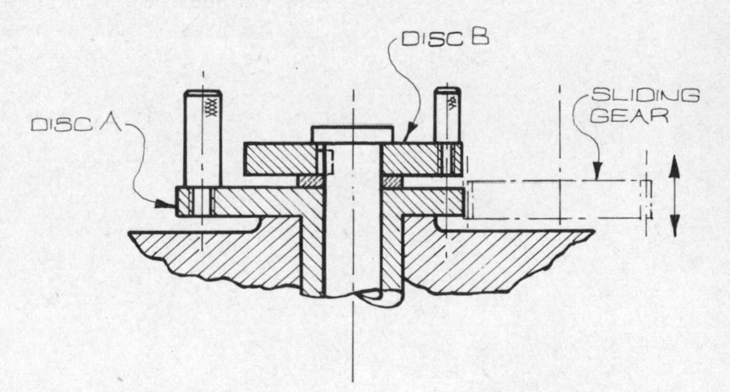

Fig. 9.20

9.3 It is required to design a suitable guard for a machine to prevent a hydraulic valve being operated until the guard is in position. Make preliminary design sketches showing alternative ways in which this can be achieved.

9.4 Construct plate cam profiles for the following requirements:

a) Base circle: $\phi 50$ mm.
Motion: Uniform velocity.
Rise: 35 mm through 180°.
Fall: 35 mm through 180°.
Follower: knife-edge.
Rotation: anticlockwise.
Line of action: through cam centre.

b) Base circle: ϕ45 mm.
Motion: 0°–180° rise 20 mm with uniform acceleration and retardation.
180°–360° fall 20 mm with S.H.M.
Follower: ϕ12 mm roller.
Rotation: clockwise.
Line of action: through cam centre.

c) Base circle: ϕ35 mm.
Motion: 0°–90° rise 36 mm with uniform acceleration and retardation.
90°–150° dwell.
150°–270° fall 12 mm with uniform velocity.
270°–360° fall 24 mm with S.H.M.
Follower: knife-edge.
Rotation: anticlockwise.
Line of action: through cam centre.

9.5 Outline the advantages and disadvantages of screw threads as devices for transmitting motion. State the common types of screw thread available and describe the particular applications of each.

9.6 Name and describe with the aid of sketches the common types of follower used with plate cams. State the advantages and disadvantages of each.

9.7 Name three types of mechanical and three types of thermal/electrical safety devices often incorporated in designs. With sketches, describe the working principles of each device and give typical applications.

9.8 Outline the common safety hazards of working machinery. In each case explain how these hazards should affect design thinking.

10 Ergonomics

10.1 Definition

Ergonomics may be defined as the scientific study of the relationship between people and their working environment. The working environment may be taken to include any factor which could affect the efficiency of the working person. Typical considerations include heat conditions, lighting, fellow colleagues, tools, machines, method and organisation of production. The object of ergonomics is thus to increase human efficiency by removing those features of design which tend to cause inefficiency or physical disability.

A full study of ergonomics would inevitably require a working knowledge of several scientific disciplines in addition to the physics and functional design. These additional disciplines include

a) *Anatomy and Physiology*—structure and function of the human body.
b) *Anthropometry*—information on body size.
c) *Psychology*—knowledge of the brain, nervous system, and human behaviour.
d) *Industrial Medicine*—this helps to define conditions of work which may prove harmful.

In this course of study, expert knowledge of these sciences will not be required, but the student will be encouraged to apply "common sense" ergonomic thought to design projects.

10.2 The Man/Machine Relationship

Any person who operates a machine may be considered as part of a closed **control-loop** system, in which he or she receives and processes information and then acts upon it. A typical control loop is shown in Fig. 10.1.

Information (e.g. the speed of a machine) is sent to the operator from a **display element** via the display communica-

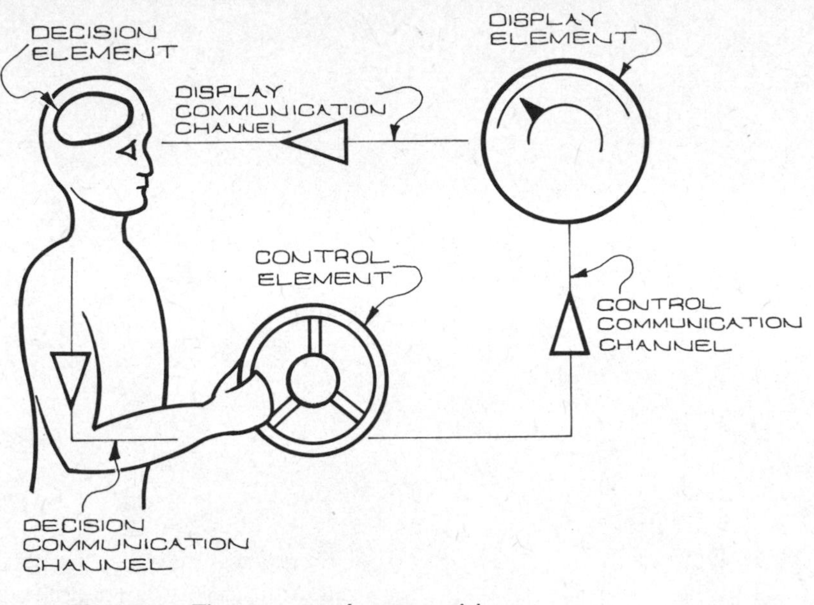

Fig. 10.1 The ergonomics control loop

tion channel. A display channel may be defined as any source of information aiding the operator in the control of the machine. Typical displays include dial gauges, digital read-outs, and warning lights.

The information from the display is passed to the control mechanism of the brain via the optical and nervous system, where it is processed to arrive at a decision in relation to required performance. The decision is then communicated to the control element via the mechanical leverage system of human bone and muscle which makes up the decision communication channel. A **control** is any device which regulates the action of a machine. Typical controls include handwheels, handles, levers, control knobs, and push-buttons. The effect on the action of the machine will be registered on the display via the control communication channel and the loop is thus completed.

The efficiency of the control loop will be affected by various internal and external factors. The display element should be easily and accurately readable and have a movement which is compatible with the movement of the control. For example, Fig. 10.2 shows obvious distinctions between compatibility and incompatibility for dials and handwheels. The position of displays should be such that they may be communicated to the operator with the minimum of physical effort.

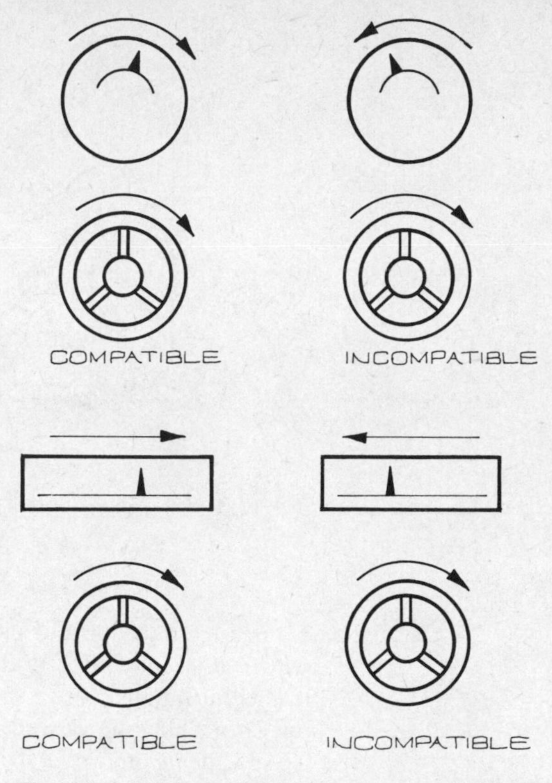

Fig. 10.2 Dial and handwheel compatibility

The display communication channel should have a direct line of path to the decision-making organs and be free from interference such as glarc on an illuminated panel.

The operator must be physically and mentally capable of making the required decision under satisfactory working conditions.

The decision communication channel should provide easy access to the control and be free from interference.

The control element should be easily operated and give compatibility with the display as already discussed.

The control communication channel must be a reliable mechanism.

External factors such as heating, lighting, noise, ventilation, physical obstructions, and fellow colleagues must also have a considerable effect on the efficiency of the control loop.

10.3 Applications of Control Loop

The ergonomics control loop may be applied to any production process involving a human being.

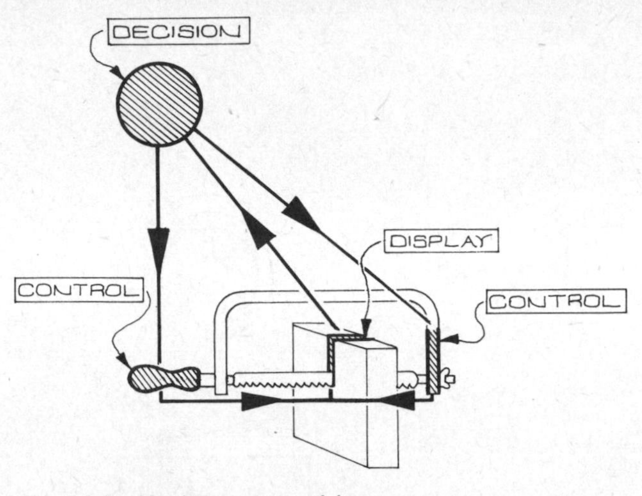

Fig. 10.3 Hacksaw control loop

1 For example, in the *process of cutting a piece of mild steel with a hacksaw* (Fig. 10.3), the display element is the quality of the cutting line which should be clearly communicated to the operator. Having viewed the cutting line, the operator will make a decision, i.e. a straight clean line will merit a decision to continue in the same manner. An unsatisfactory line will require a decision to change the angle of saw or the action of cutting. The decision is communicated to the control, which in this case is the handle of the saw and the front of the frame. The handle should be comfortable, given an efficient grip, and have a sensible size of fit for a human hand. The saw blade forms the control communication channel, whose efficiency will be affected by the design of teeth, mechanical properties of the blade material, and the tension applied to the blade.

2 In the *process of viewing a specimen through a metallurgical microscope* (Fig. 10.4), the display element is the clarity of the contour being observed. This is communicated to the viewer's eye via a lens system and electrical illumination. The lenses should be made of sufficiently clean material of high manufactured standard, and correct to the specified magnification. The illumination should be adequately bright and serviceable. Also the eyepiece should be of usable diameter and comfortable height. The display having been communicated, a decision as to the clarity is made. If unsatisfactory, this will be communicated to the control, which in this case is either a focussing knob or a knob for lateral table movement.

Separate controls for fast or fine adjustment may be incorporated. The selection of these is part of the decision-making process. These control knobs should give an easy rotational

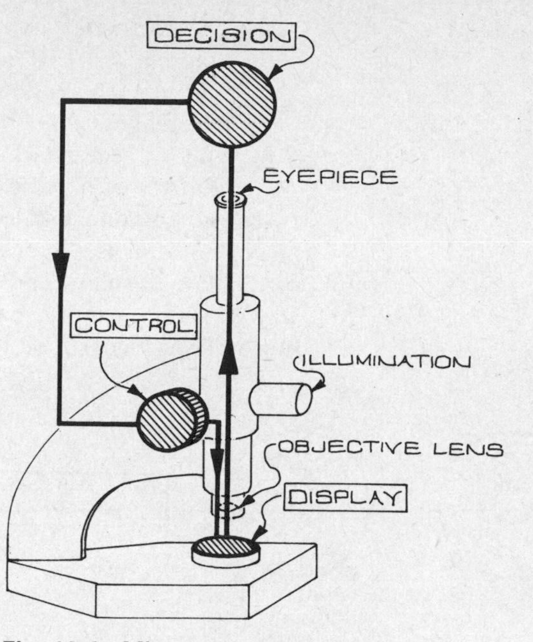

Fig. 10.4 Microscope control loop

movement, have an efficient grip, be of sensible diameter for the human thumb and forefinger, and have a sensible axial height from the work-surface to accommodate a human hand. The control communication channel takes the form of the mechanism which converts the rotational movement of the control knob into the lateral movement of the focus lens or table. This is usually achieved with a rack and pinion, whose efficiency will be affected by the design of gear profile, quality of manufacture, and choice of material. (See also p. 126.)

10.4 Practical Aspects of Ergonomics

The practical application of ergonomics is found in nearly every branch of engineering. Many engineering companies, for example the automobile industry, spend large sums of money to ensure that their product is designed to give ease of operation and comformity of shape for operator comfort.

It is the designer's task to consider the **comfort of the operator** when making layouts. In addition to those discussed in the control loop section, some examples of ergonomic consideration in design are listed below:

1. The correct position of instrument panel for ease of reading.
2. The correct position of handwheels on a machine tool.
3. The correct position of the magnification-adjusting knob on a micrometer, and its size.

4 The correct size and shape of a micrometer for ease of holding.
5 The correct positioning of handles of simple tools as on a file, plane, etc.
6 The correct relationship between the driver, seat and controls of a vehicle.

Instrumentation and control form the basics of many engineering designs. Figure 10.5 shows some common methods of display. Displays and controls should be placed in such a way that they can be read either from left to right or top to bottom(see Fig. 10.6).

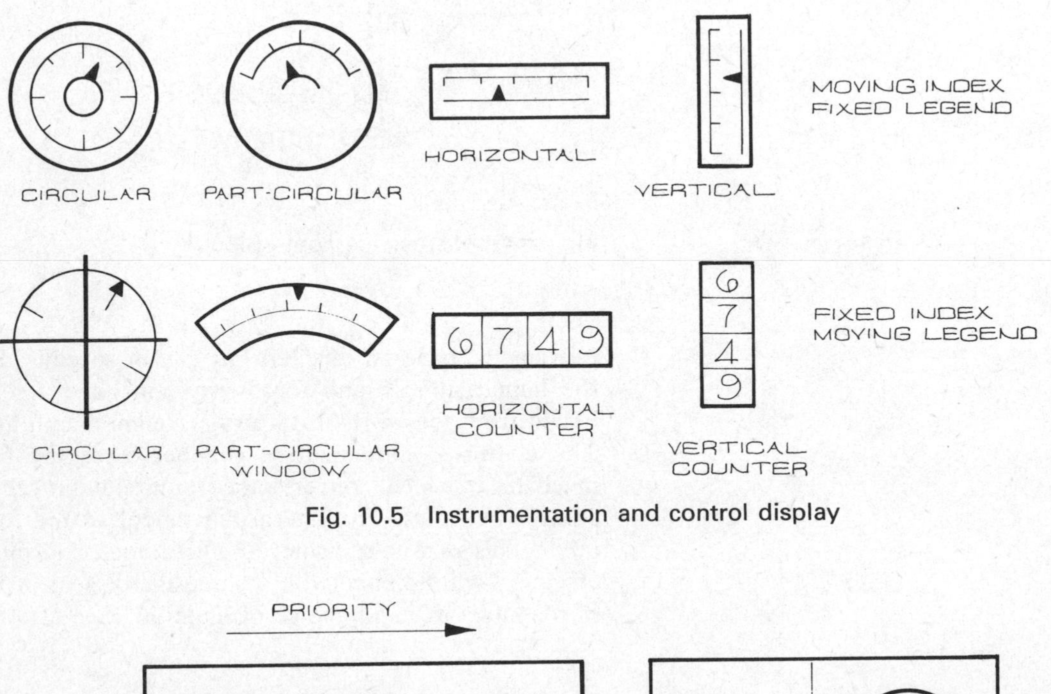

Fig. 10.5 Instrumentation and control display

Fig. 10.6 Displays and controls: order of priority

It is not essential to place the dials in such a way as to be pleasing to the eye. It is more important that they be placed in order of priority. However, if the instruments on dials have to be watched continuously, then from a health point of view, it is necessary for the designer to shape the dials in such a way as to give an acceptable visual display. Figure 10.7 shows a possible display layout for a car facia panel.

The **inspection** of components is an important operation in the manufacturing sequence. The instruments used in checking for size should be designed so that they do not place any undue strain on the inspector such that resulting fatigue causes errors in measurement.

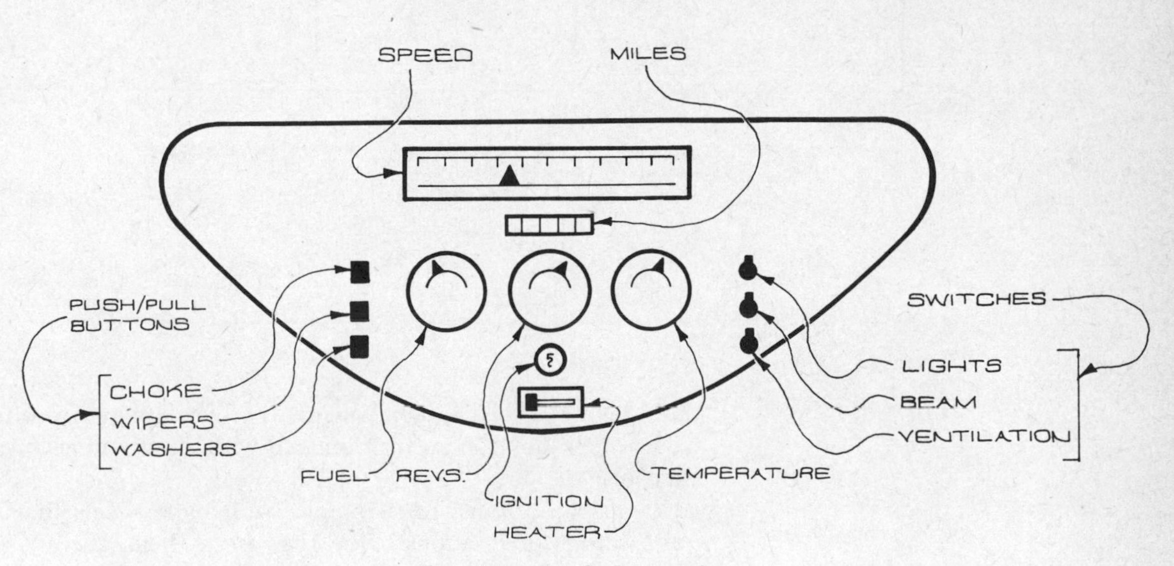

Fig. 10.7 Car facia panel

Figures 10.8 and 10.9 show the right and wrong ways of positioning a dial measuring instrument. The dial measuring instrument shown in Fig. 10.8 is positioned so that the inspector can comfortably hold the piece to be checked and read horizontally the dimension shown on the dial. In Fig. 10.9, however, the dial measuring instrument is too high for the operator to comfortably operate and results in inaccurate measuring and fatigue of the inspector. If possible an adjustable-height device should be added to the stand so that it can be set at varying heights depending on how tall the operator is.

It is *acceptable* for a component to be placed on a machine at a low level so that it necessitates the service engineer in stooping or contortion to rectify the part, but it is *essential* that

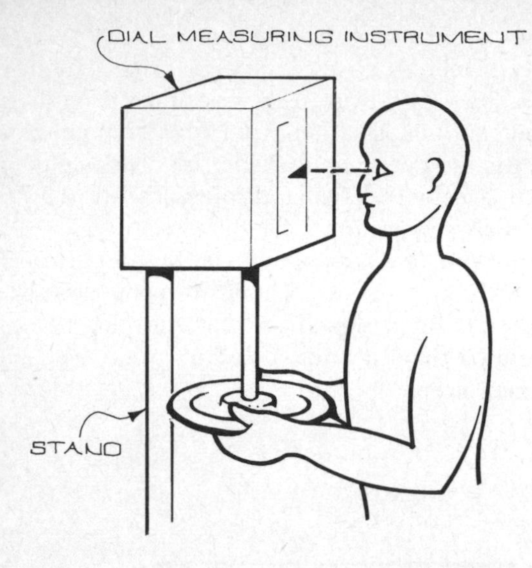

Fig. 10.8

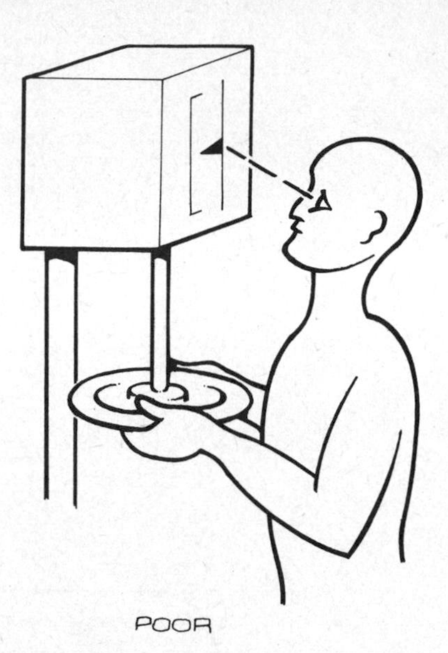

Fig. 10.9

the designer places all controls for the operator in a position from which they can be manipulated without any unnecessary exertion.

On machine tools, for example, with handwheels that require setting graduations to be read from them, the graduations should be positioned so that the operator can read them accurately. Figure 10.10 shows an illustration of how these may be read vertically and allow the operator to make an accurate setting.

10.5 The "Average" Person

Physically, people vary in size, weight, strength, shape, and their abilities to see and hear. As a yardstick, the ergonomic designer must therefore use *average* human data. Anyone with a knowledge of statistics will realize that a simple average or mean value can be very misleading and may require further analysis regarding the scatter from this value.

Figure 10.11 shows how an ironing board may be designed using women's physical data for a common range of scatter.

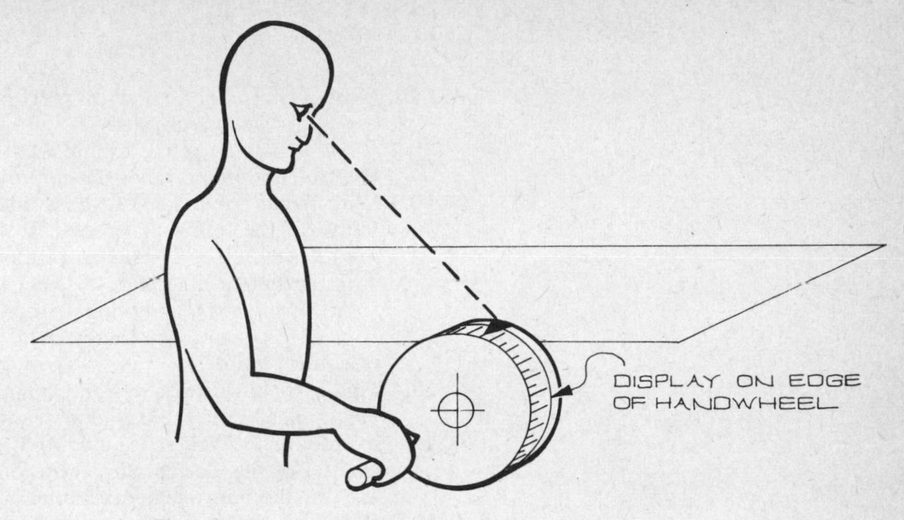

Fig. 10.10

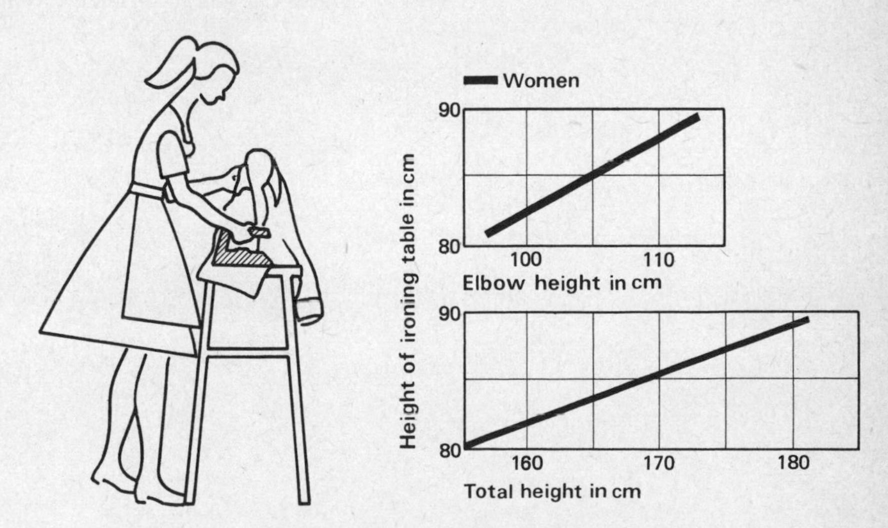

Fig. 10.11 Data used for design of ironing board (*Courtesy The Design Council*) (Data taken from Thiberg)

Problems

10.1 Give a simple definition of "ergonomics", and list the typical considerations involved.

10.2 Using a simple sketch explain what is meant by the Ergonomic Control Loop, and label the important stages in the loop.

10.3 With the aid of simple sketches, identify the ergonomic control loops for the following processes:
a) Using a wood plane to produce a flat surface.
b) Drilling a hole with an electric bench drill.
c) Machining a component on a milling machine.
In each case state the factors which could affect the efficiency of control loop.

10.4 Figure 10.12 shows a Design Council award micrometer with an electronically operated digital readout, designed and produced by Moore & Wright (Sheffield) Ltd. Compare the ergonomic aspects of this design with those of a conventional micrometer and list the improvements made.

10.5 With the aid of sketches, give typical applications of required compatibility between the control and display in working machines. (For guidance refer to Fig. 10.2, p.117.)

10.6 Produce two alternative display arrangements to the car facia panel shown in Fig. 10.7. A selection of manufacturer's handbooks will give a good guide in this exercise.

10.7 Design typical facia panels for efficient use of the following components:
a) Electric cooker
b) Centre lathe
c) Stereophonic hi-fi music centre.

10.8 Outline the major ergonomic considerations in the design of a motor cycle. Compile a list of desirable improvements in the ergonomic design of some recent models.

Fig. 10.12 Micrometer with digital readout (*Courtesy The Design Council*)

Typical examples of plastics and nylon components used in industry. (*Courtesy Polypenco Ltd.*)

This modern microscope (*above*), manufactured by Vicker's Instrument Ltd, which won the 1970 Council of Industrial Design Award, shows how practical ergonomics and good aesthetic qualities may be combined effectively (see page 118). (*Courtesy The Design Council*)

(*Top, page 127*) This speed reducer is designed and manufactured by Hansen Transmissions Ltd. Its split housing, overhung input bearing pedestal, and covers are made from grey cast iron. Reduction is achieved via spiral-bevel gear input, followed by a compound helical gear system. Taper roller bearings are used throughout. Note the aesthetic qualities of the dipstick.

(*Bottom, page 127*) This casing for a power plane shows the complexity of shape and quality of finish which may be achieved with zinc-alloy die casting. (*Courtesy Zinc Development Association*)

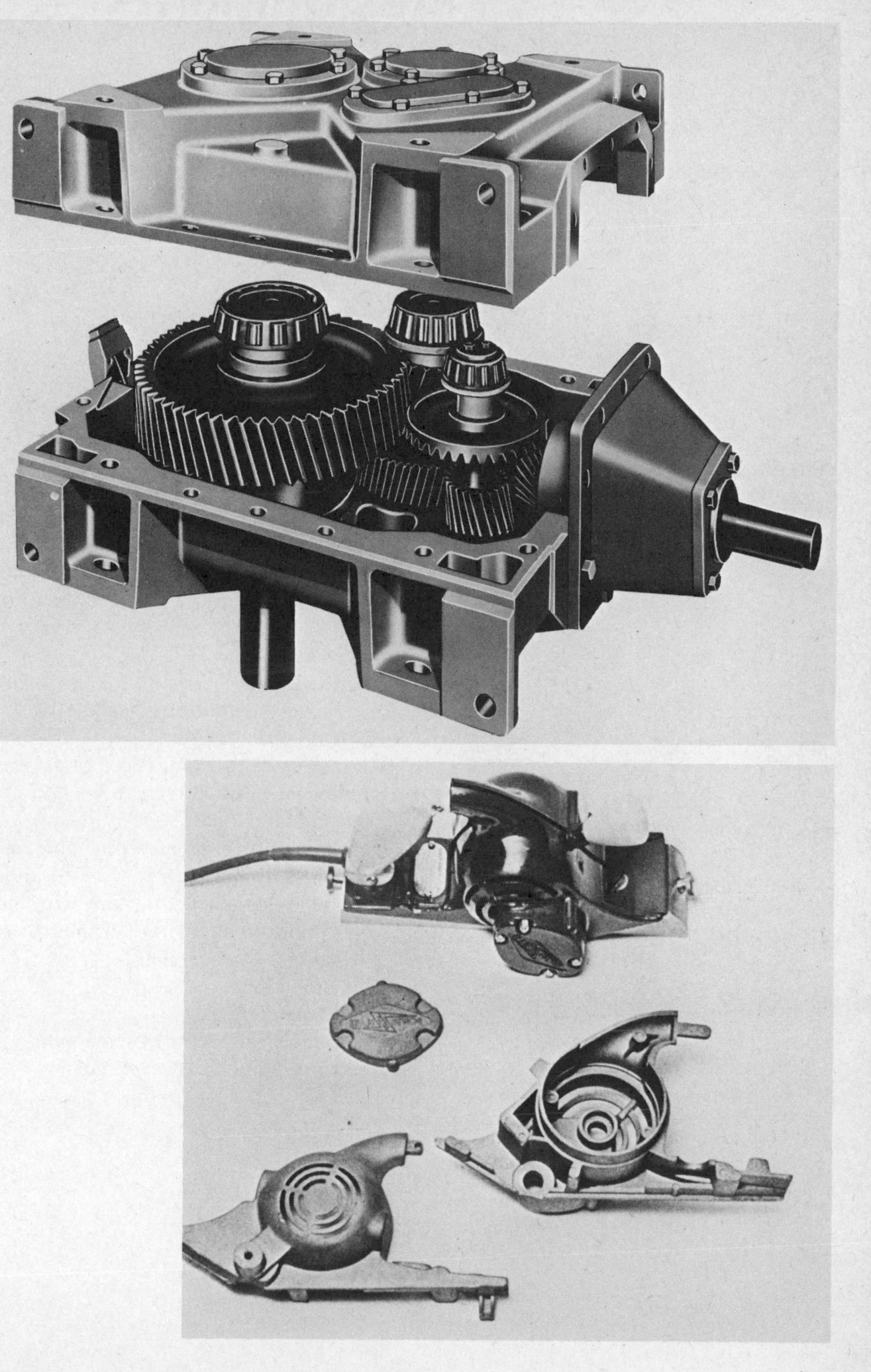

11 Assignments

Assignment 1

The Figure shows some items of a small chain-driven gear-pump for oil or suds.

Item (*a*) shows two views of the pump body.

Items (*b*), (*c*), (*d*), and (*f*) are complete views of driven gear, shaft, gear bush, and gland neck bush, respectively.

Items (*e*) and (*g*) are incomplete views of driver gear shaft and chain sprocket respectively.

1 Fill in the blank spaces of the boxes in items (*b*), (*c*) and (*d*), to indicate the following geometric tolerances in accordance with BS 308 Part 3:

a) Parallelism between vertical gear faces
b) Squareness between gear bore and gear face
c) Straightness of shaft
d) Cylindricity of bush outside diameter
e) Concentricity between bush bore and outside diameter.

2 The involute spur gears in items (*b*) and (*e*) each have 20 teeth and a module of 2.5. Determine these dimensions:

a) The P.C.D. of each gear
b) The addendum
c) The dedendum
d) The outside diameter of each gear
e) The root diameter of each gear
f) The total depth of tooth
g) The circular pitch.

Assignment 1

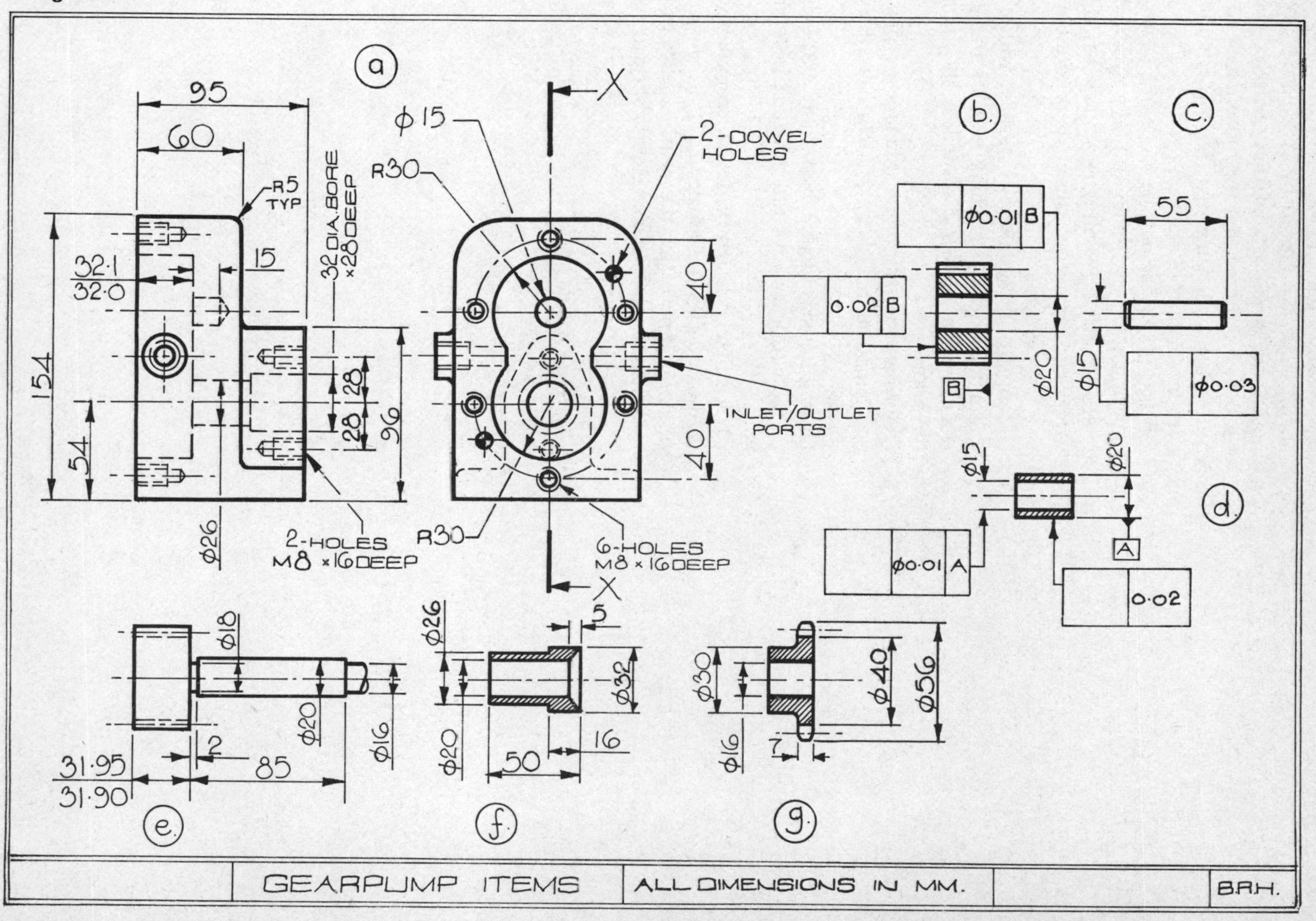

Assignment 2

The Figure shows part of the drive assembly to a gearbox input shaft, which is powered by an electric motor. Accurate alignment between motor shaft and gearbox shaft cannot be guaranteed. It is required that the motor shaft be connected to the gearbox shaft by a flange-type flexible coupling of approximate outside diameter 250 mm.

It is envisaged that the pins be made of En1A steel and have a minimum diameter of 14 mm and be positioned on a P.C.D. of 200 mm. The drive assembly is subjected to a radial torque of 1200 Nm and duty requirements necessitate a safety factor of 10.

1 Determine a suitable material, width, depth and minimum length of parallel key for the gearbox shaft, using the materials and standards sheets in the back of the book.

2 Determine the minimum number of pins required to accommodate the shear loading caused by the applied torque.

3 Draw a full-size sectional assembly of the drive including all the necessary keys and threaded fastenings, complete with

a) A suitable design of flexible coupling with flexibility method clearly shown.

b) An end cover for the gearbox with bearing location and oil-sealing device clearly shown.

Assignment 2

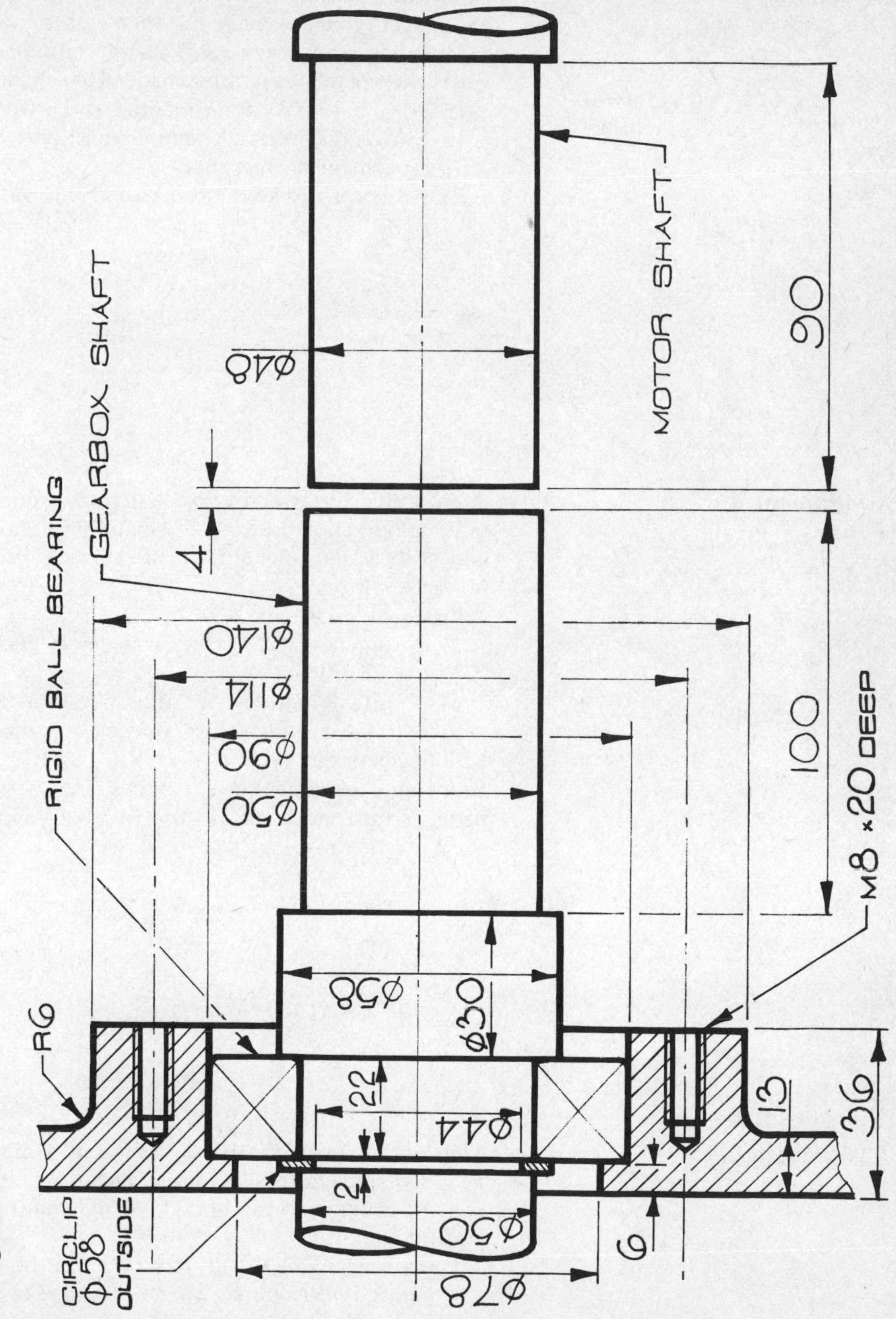

Assignment 3

The Figure shows the cast housing of a flange-mounted worm and wheel reduction unit, together with its wormshaft.

Draw a sectional general assembly on plane X–X to show

a) The completed arrangement of meshing.
b) A suitable bearing assembly for the worm shaft.
c) Two end covers accommodating lubrication sealing arrangement on one side.
d) All necessary keys, spacers and threaded fastenings.

Assignment 4

A hydraulic linear activating system is required to provide a braking operation between the hand lever shown in the Figure and the 150 mm diameter shaft which is free-wheeling after power is cut off.

Produce design schemes of

a) A suitable master cylinder assembly, including linkage to the hand lever.
b) A suitable brake-shoe, with operation provided by either two slave cylinders or one slave cylinder and linkage mechanism.

Both hand lever and brake shoes are to be provided with spring return mechanisms. The hydraulic piping system used need not be drawn.

Assignment 5

The Figure shows the basic outline of a medium-duty screw jack. Complete a feasible design arrangement of screw jack to convert rotation of the horizontal shaft into linear rise of the vertical spindle.

The arrangement should include a cast housing, covers and any required bearings, seals, and threaded fastenings.

Produce detail drawings of at least four of the main components and decide on suitable materials for each item.

Assignment 3

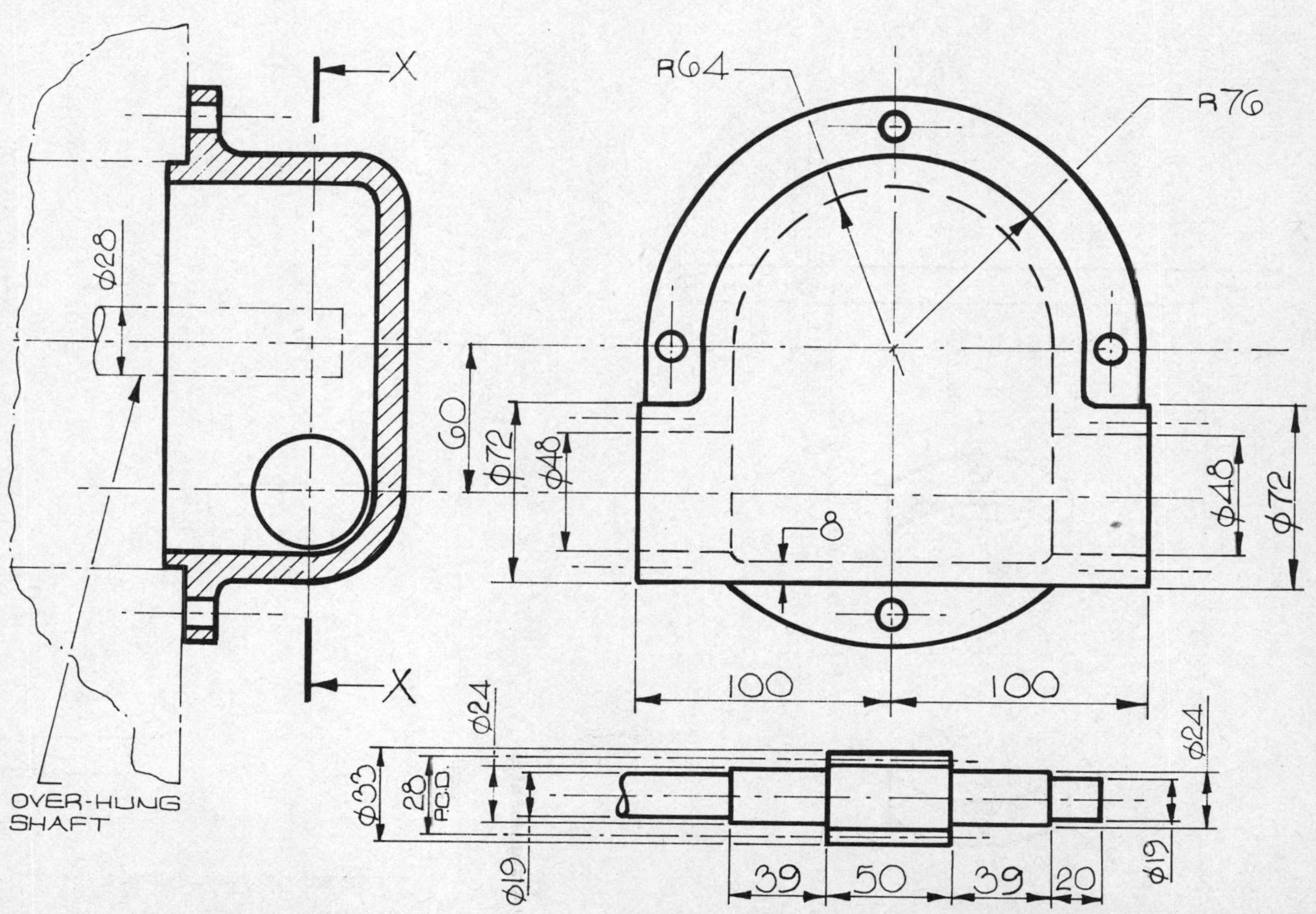

Assignment 4

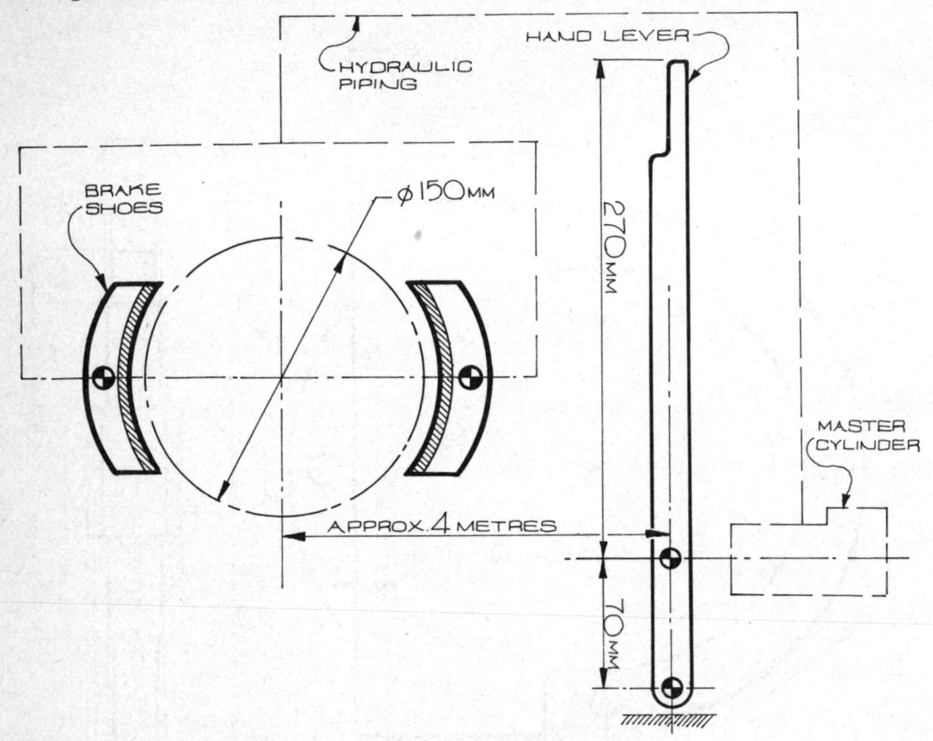

Assignment 5

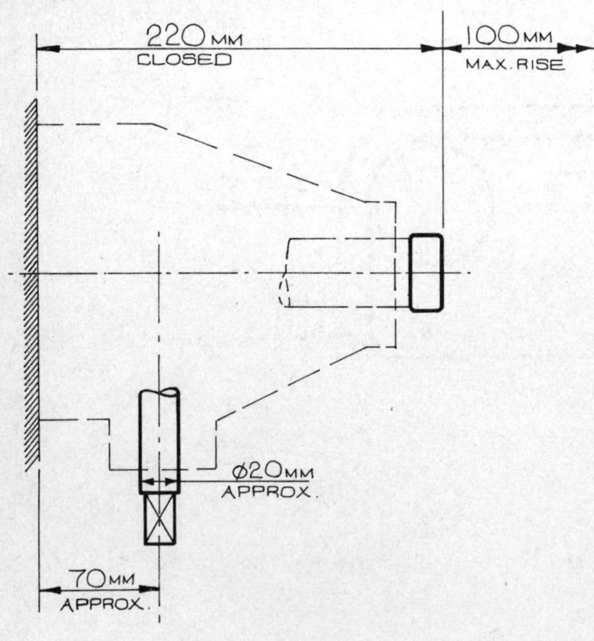

Assignment 6

Using Chapter 2 for reference, complete the following project. The Figure shows a part section of an internal combustion engine. Draw the cross-section to a suitable size by scaling, and using Charts.

Calculate

a) The gudgeon pin diameter X. The material is to be En24, and the factor of safety is 12.

b) The safety factor of the crankshaft diameter Y. Take ∅Y as 20 mm. Material is to be spheroidal graphite cast iron.

c) A suitable key size for the ∅32 mm crankshaft diameter. Material is to be En8, and the factor of safety is 12.

d) The minimum number of cylinder head studs. Material is to be En32A, and the factor of safety is 12.

e) If the number of teeth in the gear attached to the crankshaft is 22, and the number of teeth in the driven gear is 50, calculate the torque at the 50T gear.

Assignment 7

The Figure shows two shafts A and B of approximately 50 mm diameter at right angles to each other and in the same plane.

Design a mechanism connecting the two shafts such that, while shaft A rotates anticlockwise with constant velocity, shaft B will oscillate as shown through approximately 90° total angular rotation. At least three preliminary ideas should be sketched and a choice made using the points system presented in Chapter 8.

(It is envisaged that at least one idea should incorporate a cam mechanism.)

Assignment 8

The Figure shows scrap views of an electric motor drive to a gear-box input shaft. It is required that the drive is completed with a simple dog clutch assembly allowing disconnection while the shafts are stationary, by means of a yoke and lever shifting mechanism. An incomplete drawing of the fixed-clutch half is also shown. Draw

a) A complete full-size sectional assembly of the clutch showing fixed and sliding halves, together with all keys and threaded fastening.

b) A section on X–X showing suitable design of shifting mechanism including all necessary keys and threaded fastenings.

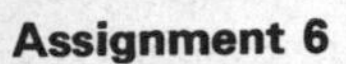

Assignment 6

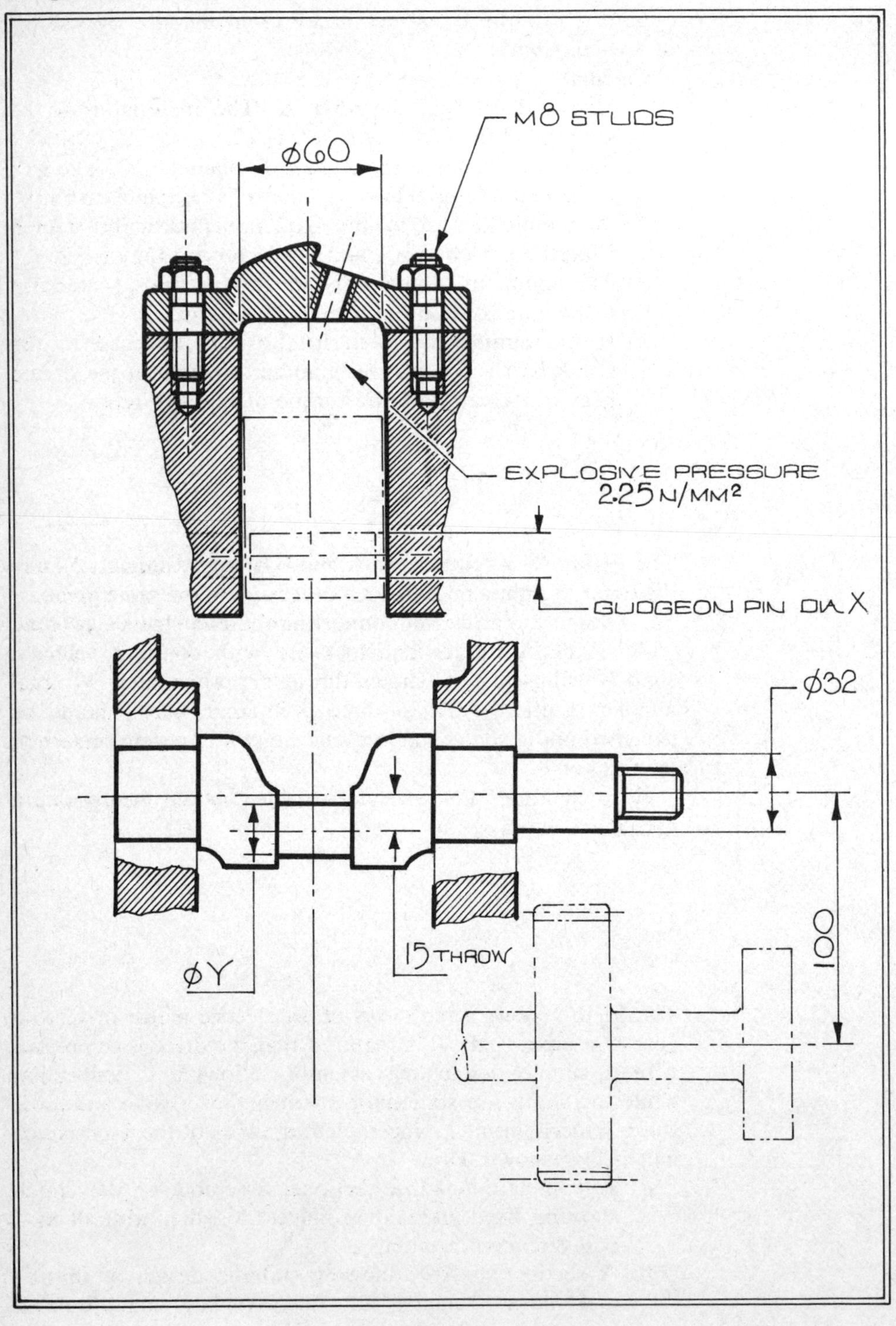

Assignment 7

Assignment 8

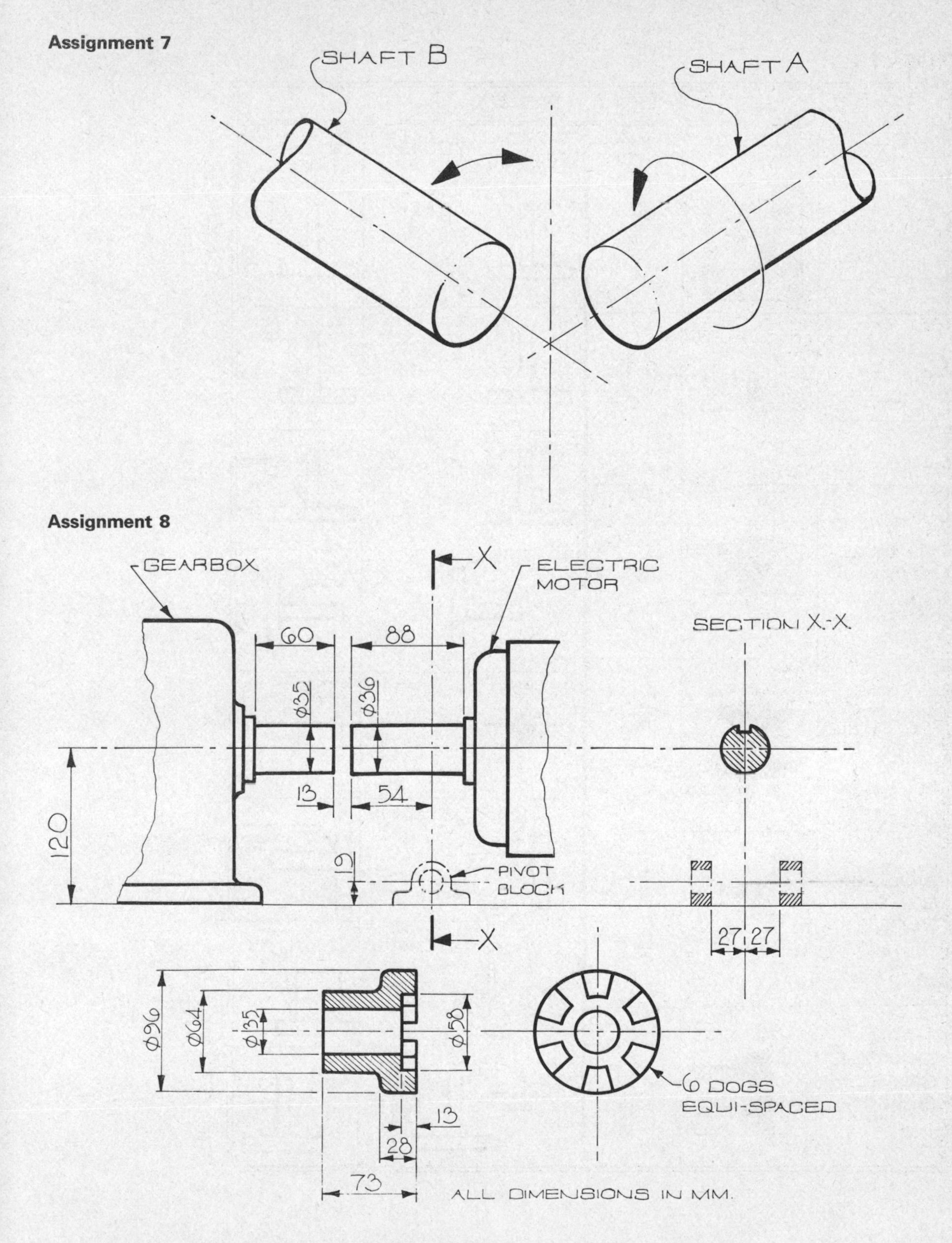

CHART 1

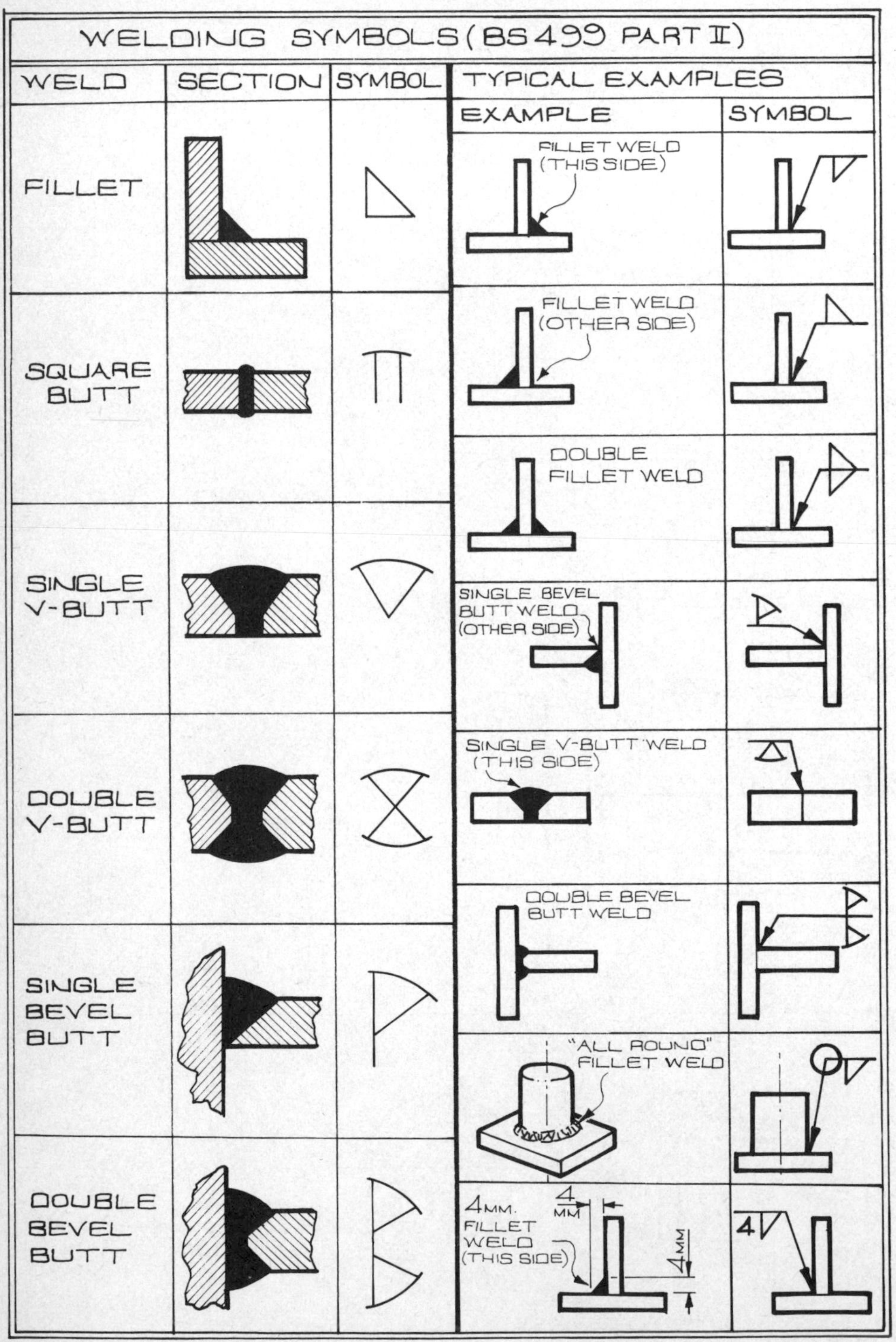
WELDING SYMBOLS (BS 499 PART II)
WELD
SECTION
SYMBOL
TYPICAL EXAMPLES
EXAMPLE
SYMBOL
FILLET
FILLET WELD (THIS SIDE)
SQUARE BUTT
FILLET WELD (OTHER SIDE)
DOUBLE FILLET WELD
SINGLE V-BUTT
SINGLE BEVEL BUTT WELD (OTHER SIDE)
DOUBLE V-BUTT
SINGLE V-BUTT WELD (THIS SIDE)
DOUBLE BEVEL BUTT WELD
SINGLE BEVEL BUTT
"ALL ROUND" FILLET WELD
DOUBLE BEVEL BUTT
4MM. FILLET WELD (THIS SIDE)
4 MM
4MM
4

CHART 2

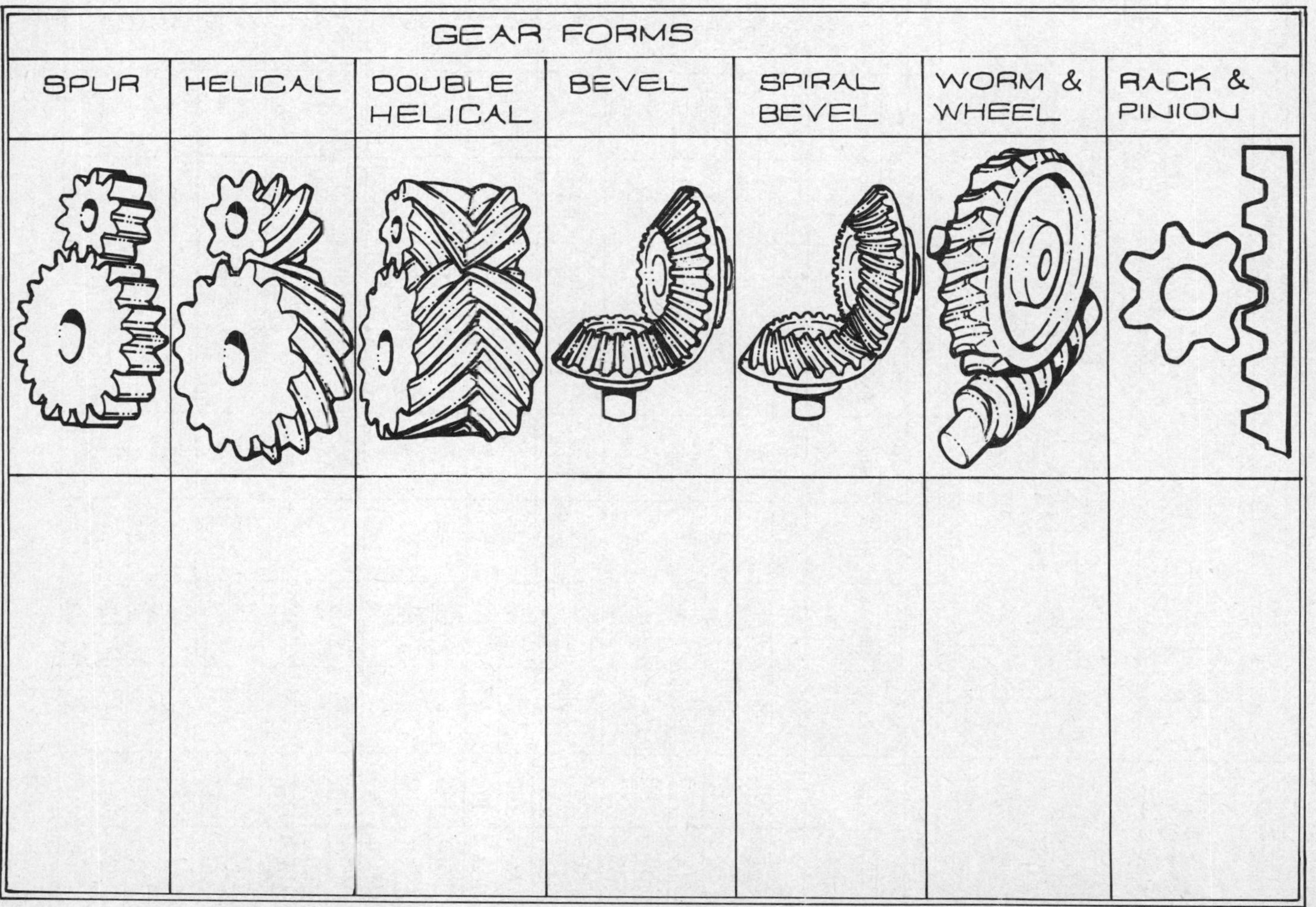
GEAR FORMS
SPUR
HELICAL
DOUBLE HELICAL
BEVEL
SPIRAL BEVEL
WORM & WHEEL
RACK & PINION

CHART 3 Selected ISO Fits: Hole Basis

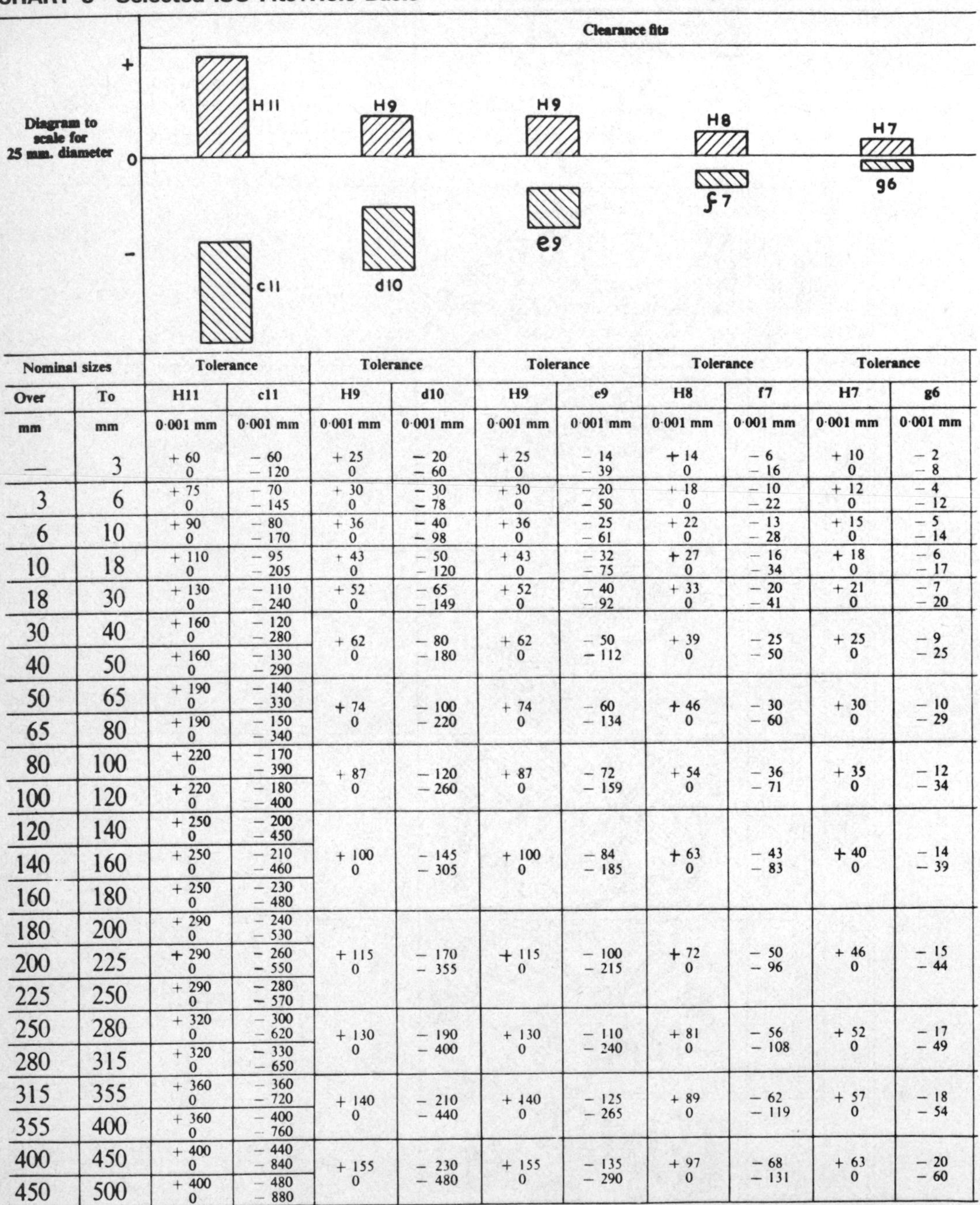

Nominal sizes		Tolerance		Tolerance		Tolerance		Tolerance		Tolerance	
Over	To	H11	c11	H9	d10	H9	e9	H8	f7	H7	g6
mm	mm	0·001 mm	0·001 mm	0·001 mm	0·001 mm	0·001 mm	0·001 mm	0·001 mm	0·001 mm	0·001 mm	0·001 mm
—	3	+ 60 0	− 60 − 120	+ 25 0	− 20 − 60	+ 25 0	− 14 − 39	+ 14 0	− 6 − 16	+ 10 0	− 2 − 8
3	6	+ 75 0	− 70 − 145	+ 30 0	− 30 − 78	+ 30 0	− 20 − 50	+ 18 0	− 10 − 22	+ 12 0	− 4 − 12
6	10	+ 90 0	− 80 − 170	+ 36 0	− 40 − 98	+ 36 0	− 25 − 61	+ 22 0	− 13 − 28	+ 15 0	− 5 − 14
10	18	+ 110 0	− 95 − 205	+ 43 0	− 50 − 120	+ 43 0	− 32 − 75	+ 27 0	− 16 − 34	+ 18 0	− 6 − 17
18	30	+ 130 0	− 110 − 240	+ 52 0	− 65 − 149	+ 52 0	− 40 − 92	+ 33 0	− 20 − 41	+ 21 0	− 7 − 20
30	40	+ 160 0	− 120 − 280	+ 62 0	− 80 − 180	+ 62 0	− 50 − 112	+ 39 0	− 25 − 50	+ 25 0	− 9 − 25
40	50	+ 160 0	− 130 − 290								
50	65	+ 190 0	− 140 − 330	+ 74 0	− 100 − 220	+ 74 0	− 60 − 134	+ 46 0	− 30 − 60	+ 30 0	− 10 − 29
65	80	+ 190 0	− 150 − 340								
80	100	+ 220 0	− 170 − 390	+ 87 0	− 120 − 260	+ 87 0	− 72 − 159	+ 54 0	− 36 − 71	+ 35 0	− 12 − 34
100	120	+ 220 0	− 180 − 400								
120	140	+ 250 0	− 200 − 450	+ 100 0	− 145 − 305	+ 100 0	− 84 − 185	+ 63 0	− 43 − 83	+ 40 0	− 14 − 39
140	160	+ 250 0	− 210 − 460								
160	180	+ 250 0	− 230 − 480								
180	200	+ 290 0	− 240 − 530	+ 115 0	− 170 − 355	+ 115 0	− 100 − 215	+ 72 0	− 50 − 96	+ 46 0	− 15 − 44
200	225	+ 290 0	− 260 − 550								
225	250	+ 290 0	− 280 − 570								
250	280	+ 320 0	− 300 − 620	+ 130 0	− 190 − 400	+ 130 0	− 110 − 240	+ 81 0	− 56 − 108	+ 52 0	− 17 − 49
280	315	+ 320 0	− 330 − 650								
315	355	+ 360 0	− 360 − 720	+ 140 0	− 210 − 440	+ 140 0	− 125 − 265	+ 89 0	− 62 − 119	+ 57 0	− 18 − 54
355	400	+ 360 0	− 400 − 760								
400	450	+ 400 0	− 440 − 840	+ 155 0	− 230 − 480	+ 155 0	− 135 − 290	+ 97 0	− 68 − 131	+ 63 0	− 20 − 60
450	500	+ 400 0	− 480 − 880								

		Transition fits				Interference fits					
H7 / h6		H7	k6	H7	n6	H7	p6	H7	s6	Holes / Shafts	
Tolerance		Tolerance		Tolerance		Tolerance		Tolerance		Nominal sizes	
H7	h6	H7	k6	H7	n6	H7	p6	H7	s6	Over	To
0·001 mm	0·001 mm	0·001 mm	0·001 mm	0·001 mm	0·001 mm	0·001 mm	0·001 mm	0·001 mm	0·001 mm	mm	mm
+ 10 0	− 6 0	+ 10 0	+ 6 + 0	+ 10 0	+ 10 + 4	+ 10 0	+ 12 + 6	+ 10 0	+ 20 + 14	—	3
+ 12 0	− 8 0	+ 12 0	+ 9 + 1	+ 12 0	+ 16 + 8	+ 12 0	+ 20 + 12	+ 12 0	+ 27 + 19	3	6
+ 15 0	− 9 0	+ 15 0	+ 10 + 1	+ 15 0	+ 19 + 10	+ 15 0	+ 24 + 15	+ 15 0	+ 32 + 23	6	10
+ 18 0	− 11 0	+ 18 0	+ 12 + 1	+ 18 0	+ 23 + 12	+ 18 0	+ 29 + 18	+ 18 0	+ 39 + 28	10	18
+ 21 0	− 13 0	+ 21 0	+ 15 + 2	+ 21 0	+ 28 + 15	+ 21 0	+ 35 + 22	+ 21 0	+ 48 + 35	18	30
+ 25 0	− 16 0	+ 25 0	+ 18 + 2	+ 25 0	+ 33 + 17	+ 25 0	+ 42 + 26	+ 25 0	+ 59 + 43	30	40
										40	50
+ 30 0	− 19 0	+ 30 0	+ 21 + 2	+ 30 0	+ 39 + 20	+ 30 0	+ 51 + 32	+ 30 0	+ 72 + 53	50	65
								+ 30 0	+ 78 + 59	65	80
+ 35 0	− 22 0	+ 35 0	+ 25 + 3	+ 35 0	+ 45 + 23	+ 35 0	+ 59 + 37	+ 35 0	+ 93 + 71	80	100
								+ 35 0	+ 101 + 79	100	120
+ 40 0	− 25 0	+ 40 0	+ 28 + 3	+ 40 0	+ 52 + 27	+ 40 0	+ 68 + 43	+ 40 0	+ 117 + 92	120	140
								+ 40 0	+ 125 + 100	140	160
								+ 40 0	+ 133 + 108	160	180
+ 46 0	− 29 0	+ 46 0	+ 33 + 4	+ 46 0	+ 60 + 31	+ 46 0	+ 79 + 50	+ 46 0	+ 151 + 122	180	200
								+ 46 0	+ 159 + 130	200	225
								+ 46 0	+ 169 + 140	225	250
+ 52 0	− 32 0	+ 52 0	+ 36 + 4	+ 52 0	+ 66 + 34	+ 52 0	+ 88 + 56	+ 52 0	+ 190 + 158	250	280
								+ 52 0	+ 202 + 170	280	315
+ 57 0	− 36 0	+ 57 0	+ 40 + 4	+ 57 0	+ 73 + 37	+ 57 0	+ 98 + 62	+ 57 0	+ 226 + 190	315	355
								+ 57 0	+ 244 + 208	355	400
+ 63 0	− 40 0	+ 63 0	+ 45 + 5	+ 63 0	+ 80 + 40	+ 63 0	+ 108 + 68	+ 63 0	+ 272 + 232	400	450
								+ 63 0	+ 292 + 252	450	500

CHART 4

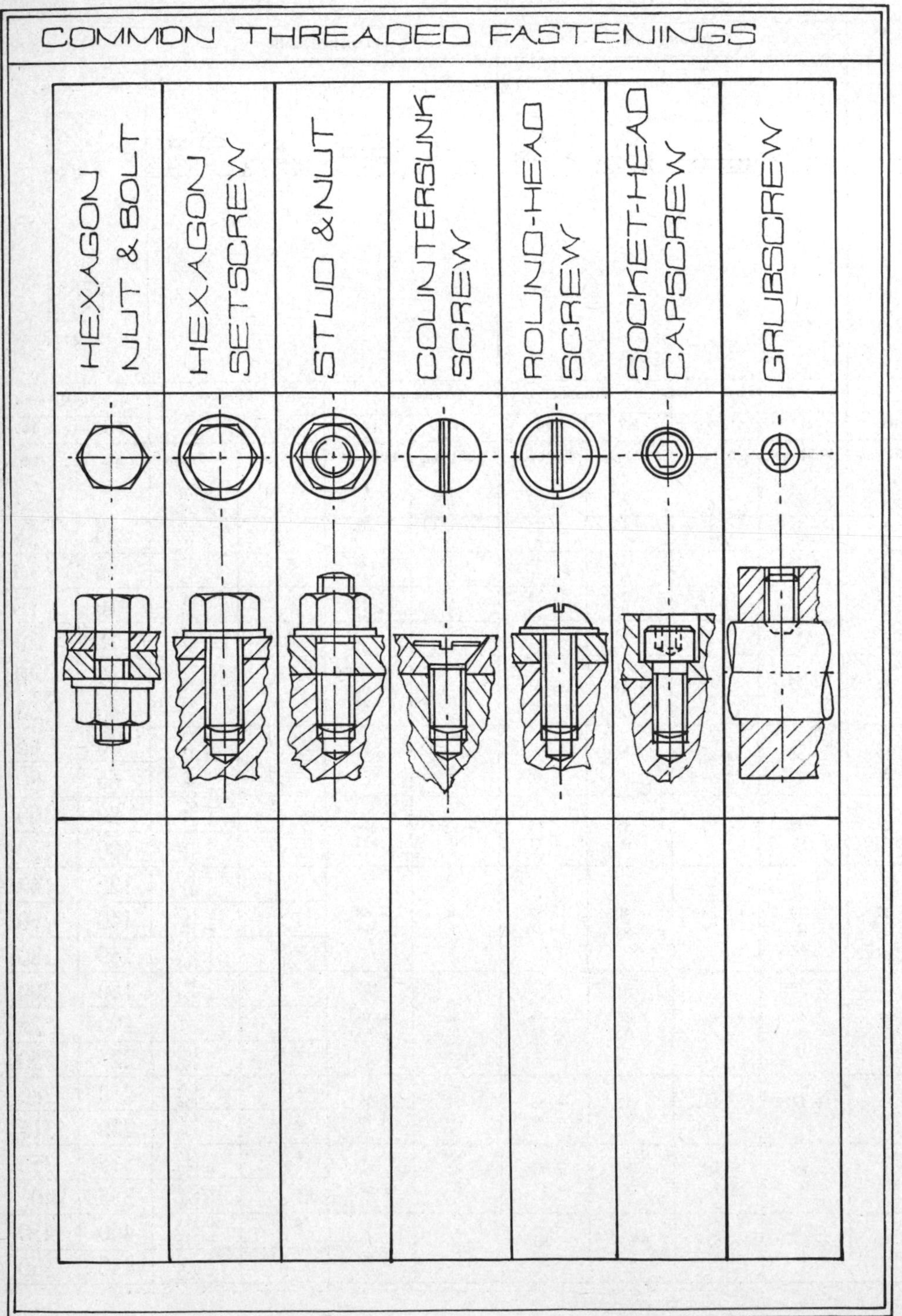
COMMON THREADED FASTENINGS
HEXAGON NUT & BOLT
HEXAGON SETSCREW
STUD & NUT
COUNTERSUNK SCREW
ROUND-HEAD SCREW
SOCKET-HEAD CAPSCREW
GRUBSCREW

CHART 5

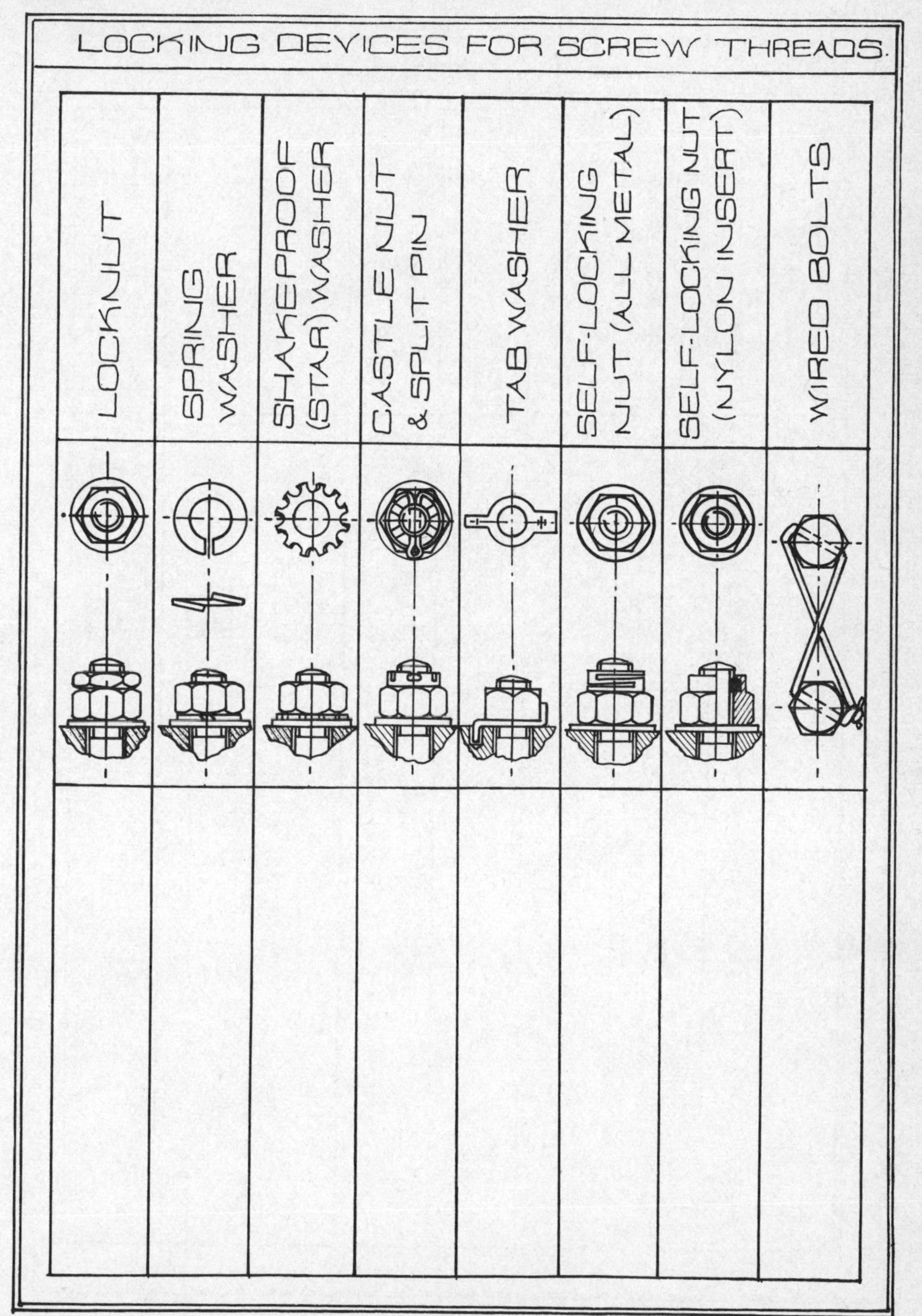
LOCKING DEVICES FOR SCREW THREADS.
LOCKNUT
SPRING WASHER
SHAKEPROOF (STAR) WASHER
CASTLE NUT & SPLIT PIN
TAB WASHER
SELF-LOCKING NUT (ALL METAL)
SELF-LOCKING NUT (NYLON INSERT)
WIRED BOLTS

CHART 6

ANTI-FRICTION BEARINGS

NEEDLE ROLLER	RIGID BALL JOURNAL	CYLINDRICAL ROLLER	BALL THRUST	TAPER ROLLER	ANGULAR CONTACT	SELF-ALIGNING BALL	SELF-ALIGNING ROLLER
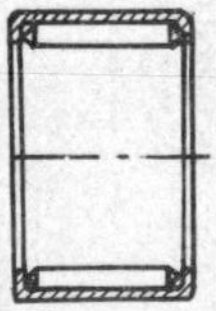	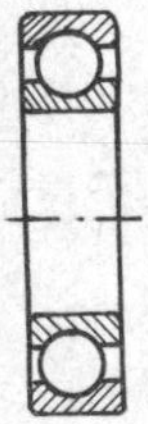	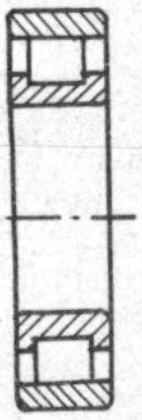		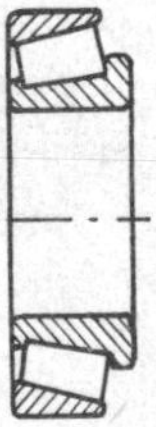	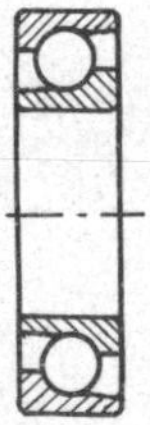	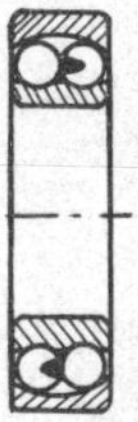	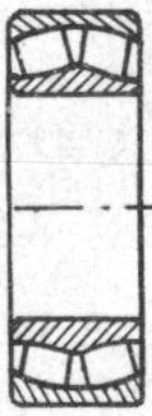

CHART 7

TYPES OF SEAL & GLAND

FELT WASHER

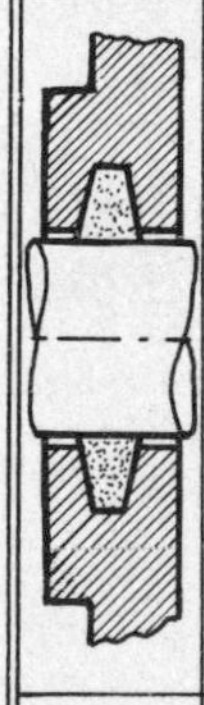

GARTER-SPRING SEAL

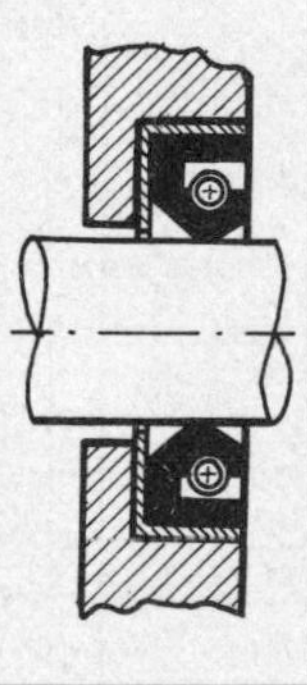

V-RING SEAL

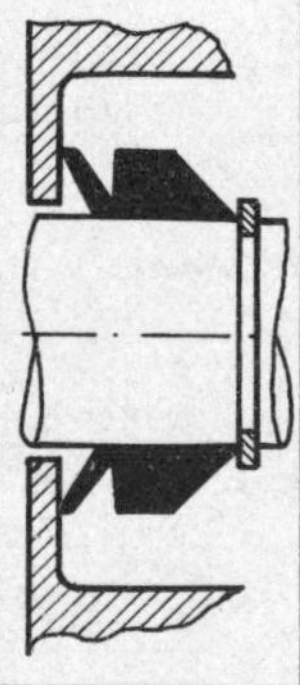

LABYRINTH SEAL

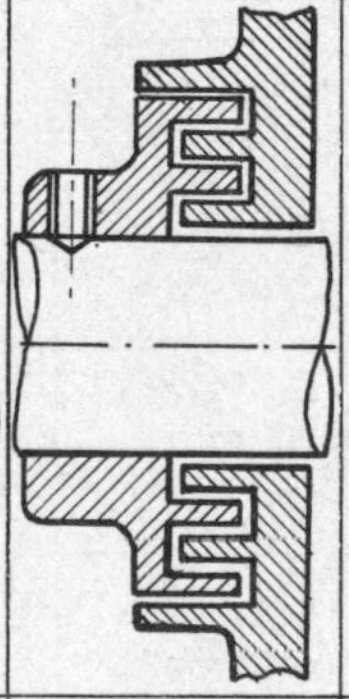

'O' RING

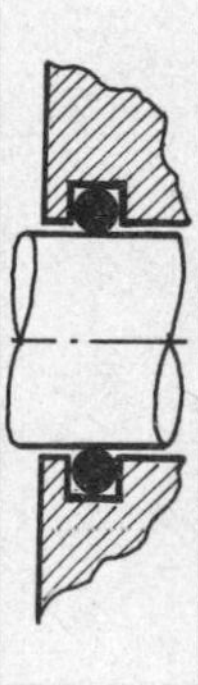

GLAND

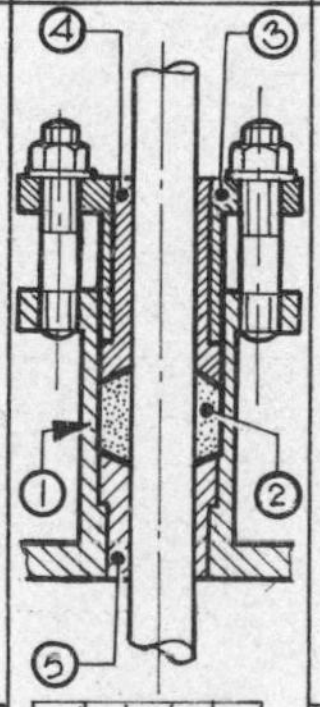

1	STUFFING BOX
2	PACKING
3	GLAND CAP
4	GLAND BUSH
5	NECK BUSH

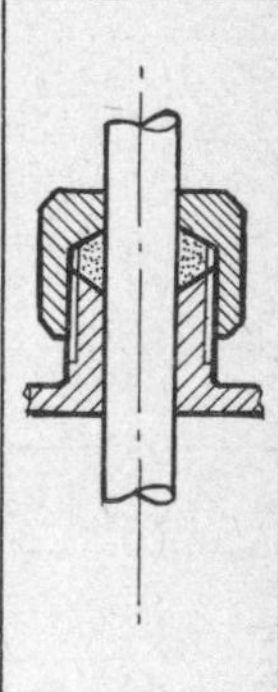

SCREWED GLAND

CHART 8

	GREY CAST IRON (GRADE 14/17)	COPPER	BRASS (60/40)	ALUMINIUM ALLOYS			ALUMINIUM
				LM14	LM6	DURALUMIN	
TENSILE STRENGTH (N/MM²)	200	150	220	280	180	400	100
COMPRESSIVE STRENGTH (N/MM²)	650	220	80	200	100	350	100
SHEAR STRENGTH (N/MM²)	110	180	110	200	120	250	60
DENSITY (kg/M³)	7200	8900	8300	2800	2600	2800	2700
BRINELL N°	150	50	65	110	70	150	40
THERMAL CONDUCTIVITY (% SILVER)	13	92	24	30	30	30	35
ELECTRICAL CONDUCTIVITY (% SILVER)	15	94	25	30	30	30	57
MELTING POINT (°C)	1200	1080	900	540	560	540	660
COEFFICIENT OF EXPANSION (/°C) $\times 10^{-6}$	11	17	20	24	24	24	24
GENERAL DESCRIPTION & USES							

CHART 9

	Carbon Steels EN42	EN32A	EN8	EN3	EN1A	Phosphur Bronze	S.G. Iron
TENSILE STRENGTH (N/MM^2)	600	500	620	460	420	350	650
COMPRESSIVE STRENGTH (N/MM^2)	600	500	620	460	420	/	650
SHEAR STRENGTH (N/MM^2)	360	300	370	270	250	200	120
DENSITY (kg/M^3)	7800	7800	7800	7800	7800	8600	7700
BRINELL Nº	200	140	200	120	110	150	200
THERMAL CONDUCTIVITY (% SILVER)	14	14	14	14	14	40	13
ELECTRICAL CONDUCTIVITY (% SILVER)	10	10	10	10	10	25	15
MELTING POINT (°C)	1400	1500	1450	1500	1500	900	1100
COEFFICIENT OF EXPANSION (/°C) $\times 10^{-6}$	12	12	12	12	12	17	11
GENERAL DESCRIPTION & USES							

CHART 10

	HIGH SPEED STEEL	CHROMIUM STEEL (EN11)	NICKEL-CHROM. STEEL (EN24)	STAINLESS STEEL (18/8)	TIN	ZINC ALLOY	NYLON (GRADE 66)
TENSILE STRENGTH (N/MM^2)	\	900	1200	600	15	200	80
COMPRESSIVE STRENGTH (N/MM^2)	\	900	1200	\	\	150	180
SHEAR STRENGTH (N/MM^2)	\	540	720	360	\	100	\
DENSITY (kg/M^3)	7900	7900	7900	7900	7300	7100	1200
BRINELL N°	660	700	340	200	5	80	25
THERMAL CONDUCTIVITY (% SILVER)	14	14	14	14	16	26	\
ELECTRICAL CONDUCTIVITY (% SILVER)	10	10	10	2	14	27	\
MELTING POINT (°C)	1400	1500	1500	1500	230	390	240
COEFFICIENT OF EXPANSION (/°C) $\times 10^{-6}$	12	12	12	16	20	27	100
GENERAL DESCRIPTION & USES							

CHART 11 ISO Metric Threads: Coarse Series

Values in millimetres except where otherwise stated

Nominal diameter			Pitch	Major diameter	Effective diameter	Minor diameter external thread	Minor diameter internal thread	Lead angle at basic effective diameter	
1st choice	2nd choice	3rd choice						deg.	min.
1.6			0.35	1.600	1.373	1.171	1.221	4	38
	1.8		0.35	1.800	1.573	1.371	1.421	4	3
2			0.4	2.000	1.740	1.509	1.567	4	11
	2.2		0.45	2.200	1.908	1.648	1.713	4	17
2.5			0.45	2.500	2.208	1.948	2.013	3	43
3			0.5	3.000	2.675	2.387	2.459	3	24
	3.5		0.6	3.500	3.110	2.764	2.850	3	31
4			0.7	4.000	3.545	3.141	3.242	3	36
	4.5		0.75	4.500	4.013	3.580	3.688	3	24
5			0.8	5.000	4.480	4.019	4.134	3	15
6			1	6.000	5.350	4.773	4.917	3	24
		7	1	7.000	6.350	5.773	5.917	2	52
8			1.25	8.000	7.188	6.466	6.647	3	10
		9	1.25	9.000	8.188	7.466	7.647	2	47
10			1.5	10.000	9.026	8.160	8.376	3	2
		11	1.5	11.000	10.026	9.160	9.376	2	44
12			1.75	12.000	10.863	9.853	10.106	2	56
	14		2	14.000	12.701	11.546	11.835	2	52
16			2	16.000	14.701	13.546	13.835	2	29
	18		2.5	18.000	16.376	14.933	15.294	2	47
20			2.5	20.000	18.376	16.933	17.294	2	29
	22		2.5	22.000	20.376	18.933	19.294	2	14
24			3	24.000	20.051	20.319	20.752	2	49
	27		3	27.000	25.051	23.319	23.752	2	11
30			3.5	30.000	27.727	25.706	26.211	2	18
	33		3.5	33.000	30.727	28.706	29.211	2	5
36			4	36.000	33.402	31.092	31.670	2	11
	39		4	39.000	36.402	34.092	34.670	2	0
42			4.5	42.000	39.077	36.479	37.129	2	6
	45		4.5	45.000	42.077	39.479	40.129	1	57
48			5	48.000	44.752	41.866	42.587	2	2
	52		5	52.000	48.752	45.866	46.587	1	52
56			5.5	56.000	52.428	49.252	50.046	1	55
	60		5.5	60.000	56.428	53.252	54.046	1	47
64			6	64.000	60.103	56.639	57.505	1	49
	68		6	68.000	64.103	60.639	61.505	1	42

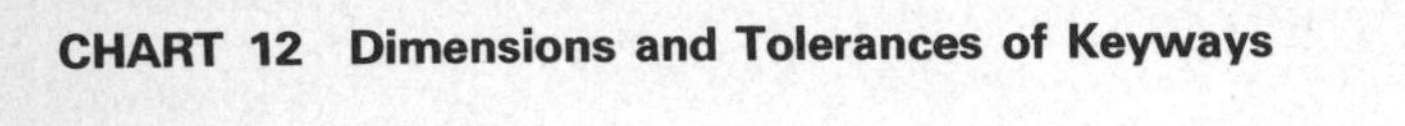

CHART 12 Dimensions and Tolerances of Keyways

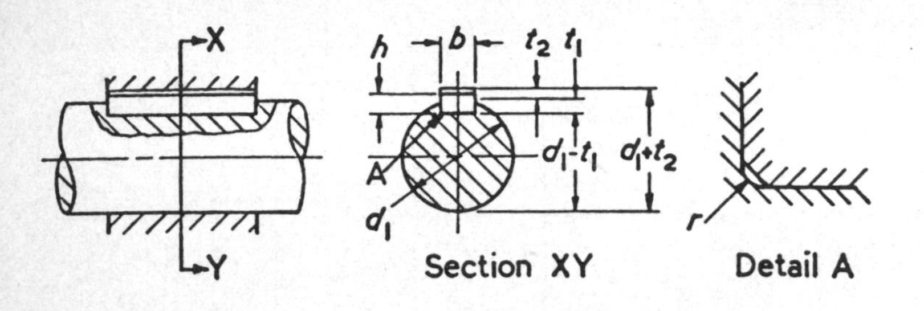

Shaft		Key (1)	Keyway											
Diameter (1) d		Section b×h	Width b						Depth (2)				Radius r	
				tolerance					shaft: t		hub: t_1			
				Sliding keys		Normal keys		Fitted keys shaft and hub						
Above	to and including		nom.	shaft H9	hub D10	shaft (N9)	hub (JS9)	(P9)	nom.	tol.	nom.	tol.	max.	min.
6	8	2× 2	2	+0.025	+0.060	−0.004	+0.0125	−0.006	1.2		1		0.16	0.08
8	10	3× 3	3	0	+0.020	−0.029	−0.0125	−0.031	1.8		1.4		0.16	0.08
10	12	4× 4	4						2.5	+0.1	1.8	+0.1	0.16	0.08
12	17	5× 5	5	+0.030 0	+0.078 +0.030	0 −0.030	+0.015 −0.015	−0.012 −0.042	3	0	2.3	0	0.25	0.16
17	22	6× 6	6						3.5		2.8		0.25	0.16
22	30	8× 7	8	+0.036	+0.098	0	+0.018	−0.015	4		3.3		0.25	0.16
30	38	10× 8	10	0	+0.040	−0.036	−0.018	−0.051	5		3.3		0.40	0.25
38	44	12× 8	12						5		3.3		0.40	0.25
44	50	14× 9	14	+0.043	+0.120	0	+0.0215	−0.018	5.5		3.8		0.40	0.25
50	58	16×10	16	0	+0.050	−0.043	−0.0215	−0.061	6	−0.2	4.3	+0.2	0.40	0.25
58	65	18×11	18						7	0	4.4	0	0.40	0.25
65	75	20×12	20						7.5		4.9		0.60	0.40
75	85	22×14	22	+0.052	+0.149	0	+0.026	−0.022	9		5.4		0.60	0.40
85	95	25×14	25	0	+0.065	−0.052	−0.026	−0.074	9		5.4		0.60	0.40
95	110	28×16	28						10		6.4		0.60	0.40
110	130	32×18	32						11		7.4		0.60	0.40
130	150	36×20	36	+0.062	+0.180	0	+0.031	−0.026	12		8.4		1.00	0.70
150	170	40×22	40	0	+0.080	−0.062	−0.031	−0.088	13		9.4		1.00	0.70
170	200	45×25	45						15		10.4		1.00	0.70
200	230	50×28	50						17	+0.3	11.4	+0.3	1.00	0.70
230	260	56×32	56						20	0	12.4	0	1.60	1.20
260	290	63×32	63	+0.074	+0.220	0	+0.037	−0.032	20		12.4		1.60	1.20
290	330	70×36	70	0	+0.100	−0.074	−0.037	−0.106	22		14.4		1.60	1.20
330	380	80×40	80						25		15.4		2.50	2.00
380	440	90×45	90	+0.087	+0.260	0	+0.0435	−0.037	28		17.4		2.50	2.00
440	500	100×50	100	0	+0.120	−0.087	−0.0435	−0.124	31		19.5		2.50	2.00

Answers to Problems

Chapter 1

1.1 36.6 kg
1.2 283.3 m
1.3 5.89 cm
1.4 15.38×10^{-6} per °C
1.5 221 W
1.7 78 233 N/mm^2
1.8 Aluminium
1.9 1.683×10^{-5}
1.10 4.348×10^{-3}

Chapter 2

2.2 *a*) 60 N/mm^2
b) EN1A
2.3 4
2.4 *a*) 9
b) 8
2.5 28.66 mm
2.6 20 mm wide × 54 mm long (EN8 material)
2.7 ϕ13.13 mm (EN8 material)
2.8 $d = 19.07$ mm, $D = 26.97$ mm, $h = 8$ mm.
2.9 6

Chapter 5

5.3 *a*) 80 mm, *b*) 4 mm, *c*) 5 mm, *d*) 70 mm, *e*) 88 mm, *f*) 12.56 mm
5.4 *a*) 75 rev/min, 60 T
b) 45 rev/min
c) Pinion: 40 mm, 17.5 mm
Wheel: 70 mm, 47.5 mm
5.6 150.35 mm
5.7

Pinion	Wheel
200 mm	350 mm
100	175
2 mm	2 mm
2.5 mm	2.5 mm
4.5 mm	4.5 mm
204 mm	354 mm
195 mm	345 mm
187.94 mm	328.89 mm
6.28 mm	6.28 mm
3.14 mm	3.14 mm

Assignments

1 (Q2) *a*) 50 mm, *b*) 2.5 mm
c) 3.125 mm, *d*) 55 mm
e) 43.75 mm, *f*) 5.625 mm
g) 7.85 mm
2 (Q2) 4
6 *a*) 8.21 mm, *b*) 12
c) 10 mm wide × 19.34 mm long
d) 5, *e*) 42 N m

Index